EARTH SCIENCE

HARPER & ROW, Publishers

New York, Evanston, and London

EARTH SCIENCE

John F. Lounsbury

Lawrence Ogden

Eastern Michigan University

PHOTO CREDITS

EARTH SCIENCE

Copyright © 1969 by John F. Lounsbury and Lawrence Ogden

Library of Congress Catalog Card Number: 69-10398

CONTENTS

Earth Science has been designed as the text for a one-semester introductory course for general education but may be used as well for an introductory course for students who plan to concentrate in the fields of geography or geology. Our objective has been to strengthen the background in earth science studies of all students, regardless of their fields of concentration, providing them with a basic knowledge of the physical earth and how it functions.

The book integrates subject matter from a number of different disciplines—geology, physical geography, meteorology, geophysics, oceanography, soil science, and astronomy—and emphasizes the interrelationships among the various fields concerned with man's physical environment. The stress is on key concepts, processes, and patterns rather than factual information for its own sake.

In organizing the material, we have avoided an overly rigid structure, thus permitting the individual instructor to shape the text to his special background and interests.

The subject matter included has been conceptualized in the course of the authors' eighteen years of experience in the teaching of Earth Sciences, General Geology, and Physical Geography at several institutions. In preliminary form, the material was used for four years for an introductory Earth Science course at Eastern Michigan University. The concepts, subject matter, and general approach of the text are those that have proved themselves most meaningful for college students at the introductory level.

J. F. L.

L. O.

PREFACE

August, 1968

EARTH SCIENCE

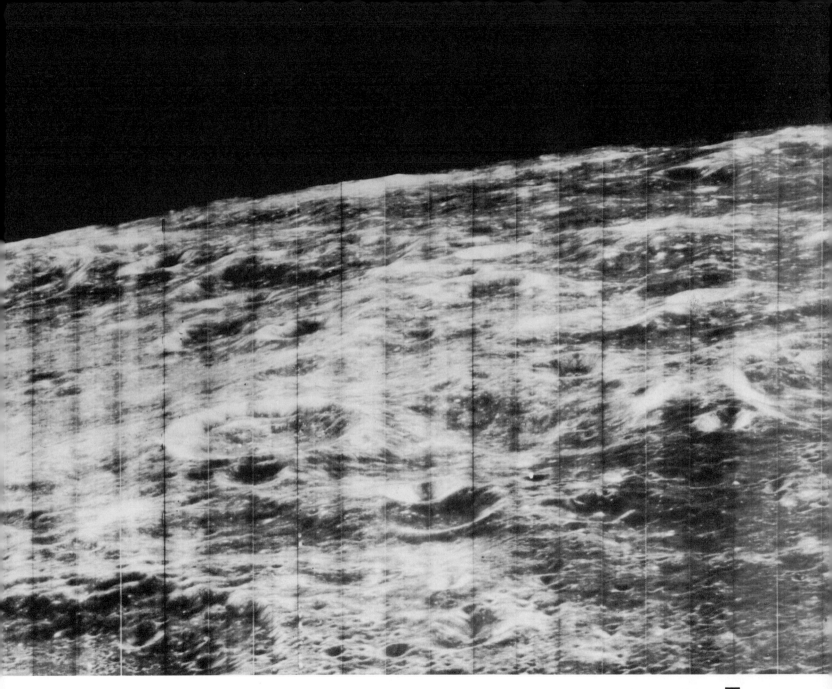

THE EARTH AS A PLANET

CHAPTER 1

On the preceding pages you see a photograph of the crescent earth, seen from about 750 miles above the back side of the moon. The picture was taken by Lunar Orbiter 1 on August 23, 1966, at 11:36 A.M. (EST), while moving away from the earth at a speed of about 3000 miles per hour—the first picture of the earth from deep space. The spacecraft was at a distance of about 232,000 miles from earth.

The picture of our planet thus seen from space, partly illuminated by sunlight and part in darkness, must be startling even to the most scientifically sophisticated observer. Nothing could illustrate more graphically what we were all taught in our early school days—that the earth is round, that it exists in space, rotating on its axis so that it is in part facing the sun and in part obscured from it.

Though it is difficult to believe, "flat earth" societies are still in existence in the United States and England, campaigning for their beliefs. And Patrick Moore reports, in his book *A Guide to the Moon*, that as recently as 1949 the British Interplanetary Society received a communication pointing out that if the earth were not flat as well as stationary, "we should otherwise be made giddy by the movement of the ground, and digestive processes would be impossible . . ."; and that ships and trains would be unable to advance if they were trying to move in opposition to the motion of a rotating earth.

Now we can all see what our planet looks like when observed from 232,000 miles in space.

THE SOLAR SYSTEM

The earth is one of nine planets that move in orbits around the sun. Some of the planets have satellites that revolve around them—as the moon circles the earth. Asteroids—small bodies less than 500 miles in diameter—as well as still smaller meteoroids and comets also revolve about the sun and form part of the solar system. Meteoroids that enter the earth's atmosphere and become luminous are known as meteors. In Table 1–1 we see the members of the solar system and the number of each presently known.

It is generally agreed that the entire solar system has a common origin, but scientists have not been able to determine with certainty which one of the several theories of its origin is correct. Perhaps interplanetary travel in the space age may provide the additional knowledge that will decide this question.

TABLE 1–1   COMPOSITION OF THE SOLAR SYSTEM

BODY	NUMBER
Sun	1
Planets	9
Satellites of planets	32
Asteroids	About 40,000 detectable on photographs
Meteoroids	Vast numbers
Comets	Vast numbers

COMPARISON OF THE PLANETS

The nine planets show marked differences when compared with respect to size, distance from the sun, and specific gravity (Table 1–2). The earth is fifth among the planets in size, with a mean diameter of 7913 miles. It is slightly flattened at the poles and bulges at the equator as a result of the centrifugal force associated with its rotation on its axis. The difference between the equatorial and polar diameters is 27 miles, giving an oblateness (polar flattening) of about 3 per cent. Note that the range in size from the earth to the largest planet is much greater than from the earth to the smallest: Whereas the diameter of Mercury is just under half that of the earth, Jupiter's is eleven times as great.

TABLE 1–2   BASIC DATA ON THE PLANETS OF THE SOLAR SYSTEM

PLANET	MEAN DISTANCE FROM SUN (millions of miles)	MEAN DIAMETER (miles)	SPECIFIC GRAVITY
Mercury	36	3,100	4.1
Venus	67	7,600	4.9
Earth	93	7,913	5.5
Mars	142	4,100	3.9
Jupiter	483	86,800	1.33
Saturn	886	71,500	0.72
Uranus	1,783	29,400	1.26
Neptune	2,794	27,000	1.61
Pluto	3,670	3,600	?

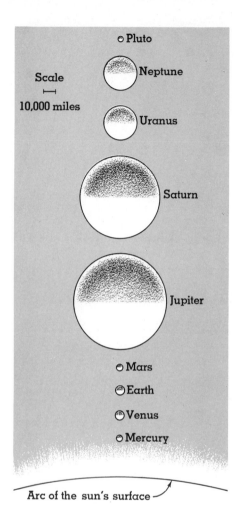

Scale
⊢━┥
10,000 miles

FIG. 1-1. Relative sizes of the planets of the solar system. The planets range in size from Mercury, less than half the diameter of earth, to Jupiter, whose diameter is approximately eleven times greater than that of earth.

The average specific gravity (density in proportion to the density of water) of the earth as a whole is about 5.5. However, the materials comprising the crust have an average specific gravity of 2.7 (continental areas) and 3.0 (oceanic areas). To give the over-all average of 5.5, densities deep in the earth must be higher, and the core may well have a specific gravity of 12 or greater. The earth's specific gravity is greater than that of any of the other planets. Saturn has a specific gravity of less than 1.0, Mercury and Venus of 4.1 and 4.9, respectively (Table 1–2).

The earth is the third planet from the sun—its mean distance is 93 million miles. This distance is taken as the definition for the unit of measurement called an *astronomical unit*. Mercury, the closest planet to the sun, has a mean distance of about 36 million miles, and Pluto, the outermost planet, 3670 million (Table 1–2). If we look at Fig. 1–1, which shows the relative sizes of the planets, and Fig. 1–2, which shows their relative distances from the sun, we see that the planets fall into two groups: the *inner planets*, which are also the smallest, and the *outer planets*, which are the largest; with the one exception that Pluto, the most distant, is next to the smallest.

THE GALAXY

Although the distances among the planets seem very great, we find that in relation to other heavenly bodies the solar system is a closely knit family. The star closest to our solar system is Alpha Centauri, which is approximately 4.15 light-years away (the velocity of light is 186,324 miles per second; the sun is 8⅓ light *minutes* from the earth). The sun is one of billions of stars that comprise the Milky Way Galaxy—a disklike aggregation of stars in which our solar system lies about ⅓ of the way from the edge (Fig. 1–3). The diameter of our galaxy is about 100,000 light-years.

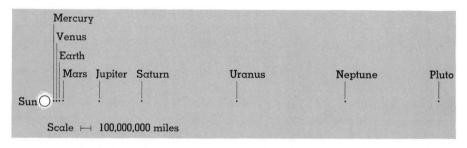

FIG. 1-2. Mean distances of the planets from the sun, shown to scale.

4

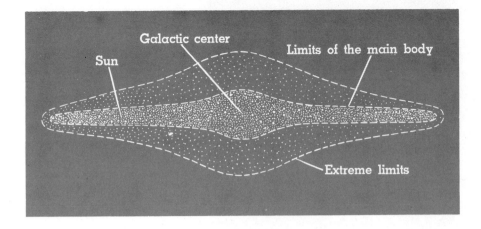

FIG. 1-3. The Milky Way Galaxy, a disk-shaped grouping of billions of stars which spreads over a diameter of approximately 100,000 light-years. The sun is located about one-third of the distance from the edge of the galaxy.

Millions of other galaxies are seen within the range of our most powerful telescopes, and it may be assumed that millions more lie beyond. Figure 1–4 shows the Great Nebula in Andromeda, the galaxy that is nearest to our own, and which astronomers believe to be close to our own in general size and shape.

MOVEMENTS OF THE EARTH

The two major movements of the earth are *rotation* (on its axis) and *revolution* (around the sun). Knowledge of these movements and the way they affect the earth's relationship to the sun is basic to an understanding of such important phenomena as the distribution of solar energy over the surface of the earth, the changing of seasons, variations in air and ocean currents, and the distribution of climates. Some of these subjects we will consider in this chapter, others will be dealt with later in the book.

Before we consider the earth's rotation and revolution, we should briefly note two other forms of motion, whose effects we do not experience in any direct way. The earth, together with the rest of the solar system, moves as part of the rotation of the galaxy as a whole. And the galaxy partakes of the expanding motion that is general throughout the universe.

ROTATION

The earth *rotates* on its axis, making one complete turn in approximately 23 hours, 56 minutes, and 4 seconds. Although the rate of rotation may be considered constant, variations in the period of rotation of up to a

FIG. 1-4. The Great Nebula in Andromeda, the galaxy that is thought to be similar to the Milky Way Galaxy in size and shape. (Mount Wilson and Palomar Observatories.)

EARTH SCIENCE

few thousandths of a second have been measured. The surface speed at the equator is about 1041 miles per hour, decreasing to zero at the axis, or poles, of the earth.

In the course of one rotation any given portion of the earth's surface faces the sun for one period of time, receiving light and solar energy, and faces away from the sun, in darkness, for another period. The length of day and night varies from one place to another and, in any location except the equator, from one season to another. At the equator, the days and nights are of about equal length throughout the year.

REVOLUTION

The earth *revolves* around the sun in an elliptical orbit. The speed of revolution is greatest when the earth is closest to the sun (perihelion) and least when the earth is farthest from the sun (aphelion). Although the average distance between the two bodies is about 93 million miles, the distance at perihelion, in January, is only about 91½ million miles, and at aphelion, in July, approximately 94½ million miles. The average speed of revolution is about 70,000 miles per hour, and the average time to complete one complete revolution in orbit about 365¼ days, or one year.

Equinoxes and solstices

The earth's axis is tilted away from the perpendicular to the plane of its orbit around the sun, and, as a result, the equatorial plane of the earth makes an angle of 23½° with the plane of the orbit. This is referred to as the *inclination* of the earth's axis. The axis of the earth at any given position in its orbit is parallel to its axis at any other position, a condition known as *parallelism*. Because of these two conditions the angle of the sun's ray at any location on the earth and the length of day and night vary from one season to another (Fig. 1–5). The angle of the sun's ray is greatest and the daylight period longest during the summer.

Lines of latitude (parallels) running true east-west around the globe are used to measure distances north and south of the equator. Latitude ranges from 0° at the equator to 90° south and 90° north at the South Pole and North Pole, respectively. Five lines of latitude are of special importance:

Equator	0°
Tropic of Cancer	23½° N.
Tropic of Capricorn	23½° S.
Arctic Circle	66½° N.
Antarctic Circle	66½° S.

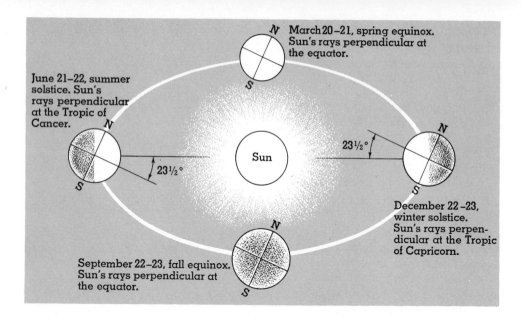

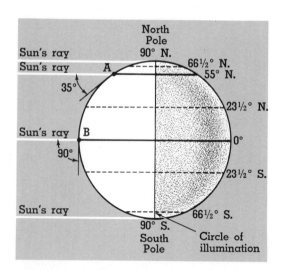

FIG. 1-6. Relationship of the sun's rays to the earth's surface during the equinoxes. The angle between the sun and the horizon is 90° at the equator (0°) (observer B); 0° at both poles (90° N. and 90° S.); and 35° at 55° N. (observer A).

TABLE 1-3   EQUINOXES AND SOLSTICES

	SUN'S RAY PERPENDICULAR		SUN'S RAY TANGENT
Equinox	Equator	March 20–21 ⎫	North Pole
Equinox	Equator	Sept. 22–23 ⎬	and South Pole
Solstice	Tropic of Cancer	June 21–22 ⎫	Arctic Circle
Solstice	Tropic of Capricorn	Dec. 22–23 ⎬	and Antarctic Circle

These lines of latitude, or north-south locations on the earth's surface, mark the relation of the ray of the sun to the earth's surface at certain times of the year called the equinoxes and the solstices; and these periods in turn mark the changes of the seasons (Table 1–3).

EQUINOXES

During one year, or one revolution of the earth around the sun, the vertical ray of the sun migrates from the Tropic of Cancer to the Tropic of Capricorn and back, crossing the equator twice. The sun's ray is directly perpendicular at the equator on March 20–21 and on September 22–23, and these are the periods of the equinoxes, vernal and autumnal, respectively (Fig. 1–6). At these times, the *circle of illumination*, or boundary between darkness and light, cuts the lines of latitude in half and, if we do not consider twilight, days and nights are of equal length over the entire earth. The sun's ray is tangent at the poles, and the circle of illumination intersects the axis of the earth at the poles.

SOLSTICES

Once each year, on June 21–22, the vertical ray of the sun is directly overhead at the Tropic of Cancer (23½° N.) and tangent at the Arctic Circle and Antarctic Circle (Fig. 1–7). In the Northern Hemisphere this is known as the summer solstice. Similarly, once each year, on December 22–23, the vertical ray of the sun is directly overhead at the Tropic of Capricorn and tangent at the Arctic and Antarctic Circles (Fig. 1–7). This is the winter solstice in the Northern Hemisphere. (The seasons are reversed in the Southern Hemisphere.) Note that the vertical ray migrates a total of 47° of latitude in the course of the year, never reaching farther poleward than 23½° north or south.

If at any given time one knows the latitude at which the sun is directly overhead, it is easy to determine the angle of the noonday sun at any degree of latitude. Simply find the number of degrees of latitude between the given location and the latitude of the vertical sun, then subtract from 90° to get the angle of the sun at that location. For example, find the angle of the noonday sun at 50° north at the winter solstice:

Degrees of latitude between 50° N. and overhead sun (23½° S.) = 73½°
$$90° - 73½° = 16½°$$

Thus, the angle of the noonday sun at 50° N. is 16½° above the southern horizon. At the winter solstice, the difference in degrees of latitude would be 47° less and the angle of the noonday sun therefore 47° greater.

FIG. 1-7. Relationship of the sun's rays to the surface of the earth during the solstices.

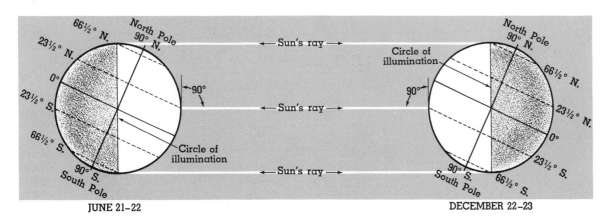

JUNE 21–22   DECEMBER 22–23

Not only is the angle of the sun's ray greater during the summer at any given location, but the period of daylight is longer and, of course, the nights shorter. During the summer in the Northern Hemisphere the days and nights are almost equal near the equator, but the daylight period increases, moving poleward, to 24 hours at the Arctic Circle and 6 months at the North Pole (Table 1–4). Because, other factors being equal, the amount of solar energy received is proportional to the length of daylight, it is possible to grow some crops in far northern latitudes in the very long days of the brief summer season.

SOLAR TIME

Solar time is measured by the apparent motion of the sun. Thus, noon on any day is calculated as the instant the sun reaches its zenith. Since the earth does not revolve around the sun at a constant speed, the time interval from one noon to the next varies slightly in the course of a year. For convenience an average is used, and this is called the *mean solar day*. The mean solar day is 24 hours—the average time it takes for the earth to make one complete rotation in relation to the sun's position.

You can see that solar time—the time according to the position of the sun—will be the same only at locations along the same line of longitude, or east-west position, and will be different at different lines of longitude (meridians). Until about 100 years ago, each locality used its own solar time and noon was calculated as the time the sun was at its zenith over the local meridian.

TABLE 1–4 LENGTH OF DAYLIGHT PERIOD IN THE NORTHERN HEMISPHERE

LATITUDE	MARCH 20–21	JUNE 21–22	SEPTEMBER 22–23	DECEMBER 22–23
0°	12 hr	12 hr	12 hr	12 hr
10°	12 hr	12 hr 35 min	12 hr	11 hr 25 min
20°	12 hr	13 hr 12 min	12 hr	10 hr 48 min
30°	12 hr	13 hr 56 min	12 hr	10 hr 4 min
40°	12 hr	14 hr 52 min	12 hr	9 hr 8 min
50°	12 hr	16 hr 18 min	12 hr	7 hr 42 min
60°	12 hr	18 hr 27 min	12 hr	5 hr 33 min
70°	12 hr	2 months	12 hr	0 hr 0 min
80°	12 hr	4 months	12 hr	0 hr 0 min
90°	12 hr	6 months	12 hr	0 hr 0 min

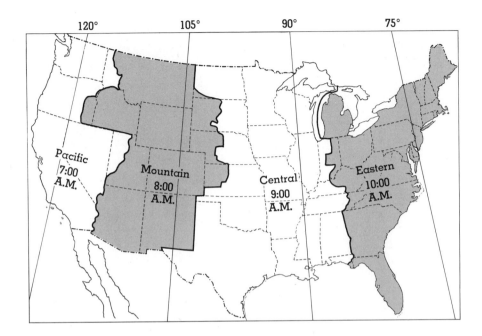

FIG. 1-8. Standard time zones in the United States.

STANDARD TIME

With the development of rapid transportation and communication, the system of local solar times became completely impractical, so that by now a standard time system has been adopted over most of the world. According to the standard time system:

1. There are 24 time zones, each running from pole to pole and each 15° of longitude in width.
2. All locations within one zone maintain the same time as the solar time of the meridian passing through the center of the zone.
3. Changes of time occur at the zone boundaries.
4. Clocks are set back one hour, moving westward, at each zone boundary.

This system is modified within the United States to the extent that zone boundaries are often adjusted to coincide with state, county, or city lines (Fig. 1–8). In addition, a few countries in the world have time zones that are not based on widths of 15° of latitude.

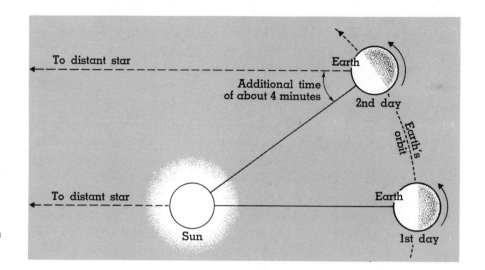

FIG. 1-9. The sidereal day is about 4 minutes shorter than the solar day owing to the movement of the earth around the sun.

SIDEREAL TIME

Sidereal time is time measured with respect to the stars. Thus, a sidereal day is the time it takes for the earth to make one complete rotation of 360° as measured by the return of a given star to an identical position in the sky. The sidereal day is about 23 hours, 56 minutes, and 4 seconds (Fig. 1–9). We have previously noted that solar time is affected by the movement of the earth in its orbit around the sun. But the distance between a star and the earth is so great that the distance traveled by the earth in its orbit in one day is infinitesimal in its effect.

INTERNATIONAL DATE LINE

Since there are 24 standard time zones, a traveler proceeding westward around the earth would make a total of 24 one-hour changes. As a result, he would find himself, on arrival back at his starting point, one day behind those he had left. To avoid this impractical consequence, it was necessary to establish a line where corrections would be made in the calendar. The International Date Line was established at the 180-degree meridian to serve this function. Some partial adjustments of the line were made to prevent inconvenience in populated areas crossed by the 180th meridian. Since the line is in the center of a time zone, there is a calendar change only and no adjustment of the clock.

For westward travelers, the calendar moves ahead one day at the International Date Line, and for eastward travelers it moves back one day (Fig. 1–10). This compensates for the 24 one-hour time changes around the globe. You can see, of course, that any specific traveler must lose or gain a full day, in spite of the fact that he may only have moved through one or two hours of time change. At all times except the instant when it is noon at Greenwich, England (longitude 0°) and midnight at longitude 180°, there are two calendar days in effect in the world.

As a practical matter, the calendar correction on boats crossing the date line is not made at the moment of crossing but at night, in order to avoid having the day divided into two dates. In effect, one full calendar day is either eliminated or repeated. Thus, in crossing westward Tuesday becomes Thursday. In crossing eastward, the correction is made by repeating Tuesday.

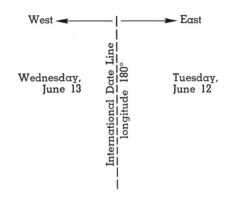

FIG. 1-10.
The International Date Line.

MOON-EARTH-SUN RELATIONSHIPS

The moon is the earth's only natural satellite (Fig. 1–11). Among the other planets that have satellites, the number presently known ranges from 2 to 12 (Table 1–5). Though not the largest of the 32 satellites in actual size, the moon is the largest in relation to the size of its own planet. It is, of course, smaller than the earth, with a diameter of 2160 miles compared with 7913 and a mass about 1/81 that of the earth. As a result of the difference in diameter, an object on the moon's surface is much closer to the center than an object on the earth's surface. Because of the combined effect of the much greater mass of the earth, which increases gravitational force, and the moderately greater size, which lessens gravitational attraction at the surface, the surface gravity of the moon is 1/6 that of the earth.

The moon is near enough to the earth so that its distance can be measured by *triangulation*, a method of computing distance by using two widely separated points on earth to get a *base line* and measuring the angles of the triangle formed by these two points and the point to be located. By this method, the mean distance between the moon and the earth is found to be about 240,000 miles. The relative closeness of the moon and its relatively large size are responsible for the behavior of tides and for many biological phenomena on earth.

TABLE 1-5

NUMBER OF KNOWN SATELLITES
FOR EACH PLANET

PLANET	NUMBER OF SATELLITES
Mercury	0
Venus	0
Earth	1
Mars	2
Jupiter	12
Saturn	9
Uranus	5
Neptune	2
Pluto	0

FIG. 1-11. The moon, the earth's only natural satellite, about one-fourth the earth in diameter. The darker areas, known as "mares" or "seas," are of lower elevation and plainlike topography. The craters are thought to have resulted from bombardment by meteors. (Mount Wilson and Palomar Observatories.)

MOTIONS OF THE MOON

The moon revolves once in its orbit around the earth in 27⅓ days with respect to a given star. This period is known as the sidereal month. During this time, however, the earth is moving in its orbit around the sun. Thus, in order for the moon to regain its original position in relation to the earth, let us say from new moon to new moon, it must travel an additional distance requiring a little more than two days. In consequence, the period of one complete revolution of the moon as viewed from the earth is about 29½ days, which is called the *synodic* month. Figure 1–12 illustrates these earth-moon relationships.

The moon rotates on its axis once in 27⅓ days—the same time it takes to make one complete revolution in its orbit. The result is that the same side of the moon is always facing the earth. (Now, space vehicles orbiting the moon can photograph the portion of its surface we have never seen.) The effect is the same as for an observer standing inside a circular automobile track and watching an automobile make a circuit—following the same side at all times.

The moon revolves around the earth in the same direction as the earth rotates on its axis. But because the earth rotates faster than the moon revolves, the moon rises in the east and sets in the west. The moon makes one complete rotation (360°) in 27⅓ days, or 13° per day. Since the earth rotates once in 24 hours, an additional 50 minutes is required for the earth to turn the 13° the moon has traveled in its orbit (Fig. 1–13). Thus, the moon rises and sets about 50 minutes later each evening.

PHASES OF THE MOON

As the moon revolves around the earth its apparent shape changes from a thin crescent to a circle and back again to a thin crescent. This occurs because the moon has no light of its own but shines by reflected sunlight. As its position in revolution with respect to the sun and the earth changes, its apparent shape changes (Fig. 1–14).

When the moon is between the earth and the sun, the side of the moon

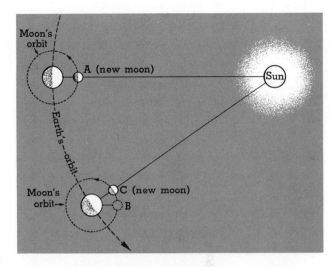

FIG. 1-12. As viewed from outside the solar system, the moon revolves around the earth (from **A** to **B** with respect to a distant star) in 27 1/3 days. During this time the earth moves in its orbit around the sun. Therefore, the moon must make more than one complete revolution from new moon to new moon. This takes about 29 1/2 days (from **A** to **C** with respect to the sun).

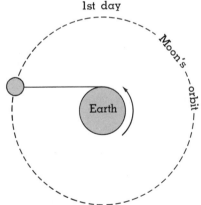

1st day

Horizon is overtaking moon and moon is rising.

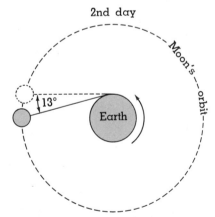

2nd day

Moon rises about 50 minutes later—time it takes for the earth to travel the additional 13° of the moon's orbit.

FIG. 1-13. The moon rises 50 minutes later each day because it revolves around the earth 13° per day, in the same direction as the earth rotates on its axis, while the earth is making one complete rotation.

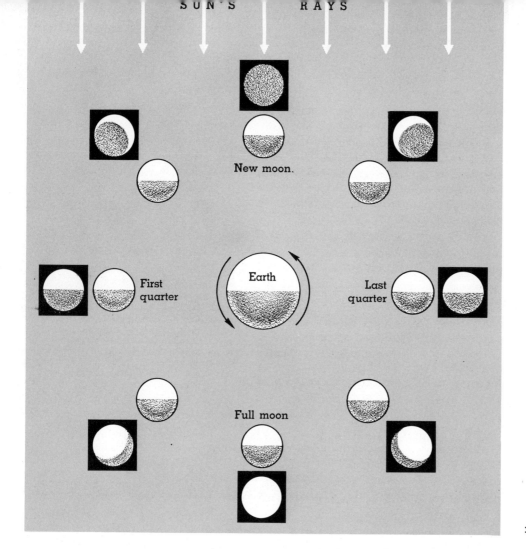

SUN'S RAYS

New moon.

First quarter

Earth

Last quarter

Full moon

FIG. 1-14. Phases of the moon.

that receives the light is the one facing away from the earth. This is called the new moon phase—the phase when the moon cannot be seen from the earth at all. Two or three days later, as the moon travels in its orbit, a thin crescent becomes visible, and then, with further revolution, the full moon. The visible portion of the moon then becomes gradually smaller, until we have the next new moon. Tides, which will be discussed in a later chapter, are closely associated with the movements that cause the phases of the moon.

16

FIG. 1-15. The total solar eclipse of November 12, 1966, observed at Pulacayo, Bolivia. The photograph shows the solar corona, or rim of the sun. An over-exposed image of Venus appears at the top. (High Altitude Observatory.)

THE EARTH AS A PLANET

FIG. 1-16. Phases of a lunar eclipse, showing the progression of the earth's shadow on the moon's surface. The picture on the right was taken shortly before total eclipse. (Yerkes Observatory.)

ECLIPSES

An eclipse occurs when the light reaching the earth from the sun either directly or by reflection from the moon is temporarily cut off as a result of the relative positions of the moon, earth, and sun. An eclipse of the sun takes place when the moon comes between the earth and the sun (new moon phase) and the moon's shadow is cast on the earth (Fig. 1–15). A lunar eclipse takes place when the earth comes between the sun and the moon (full moon phase) and the earth's shadow is cast on the moon (Fig. 1–16).

In a total eclipse of the sun, the shadow the moon casts on the earth may be up to 167 miles in width, and the path of the shadow may cross the earth's surface in a linear distance of up to 8000 miles. A solar eclipse is thus visible from only certain parts of the earth at any one time. A lunar eclipse is visible to all observers on the side of earth facing the moon. Both solar and lunar eclipses may be either total or partial, depending upon the relative positions of the earth, sun, and moon.

The moon's orbit is inclined to the earth's orbit at an angle of a little more than 5°. Because of this inclination, the moon's shadow usually falls below or above the line from earth to sun during the new moon period and the earth's shadow below or above the moon during the full moon period. If the moon's orbit were in the same plane as the earth's, the three bodies would be in line twice a month, and the result would be a solar eclipse every new moon and a lunar eclipse every full moon. As it is, the greatest number of solar eclipses possible per year is five and the least two; the greatest number of lunar eclipses is three, and there may be none.

SUGGESTED REFERENCES ——————————————

Abell, G.: *Exploration of the Universe*, Holt, Rinehart and Winston, New York, 1964.

Fath, E. A.: *Elements of Astronomy*, McGraw-Hill Book Company, Inc., New York, 1956.

Krogdahl, W. S.: *The Astronomical Universe*, The Macmillan Company, New York, 1962.

McIntyre, M. P.: *Physical Geography*, Ronald Press Company, New York, 1966.

Spar, J.: *Earth, Sea, and Air*, Addison-Wesley Publishing Company, Inc., Reading, Mass., 1965.

Strahler, A. N.: *The Earth Sciences*, Harper & Row, Publishers, Inc., New York, 1963.

Trewartha, G. T., A. H. Robinson, and E. H. Hammond: *Elements of Geography*, McGraw-Hill Book Company, Inc., New York, 1967.

Van Riper, J. E.: *Man's Physical World*, McGraw-Hill Book Company, Inc., New York, 1962.

Wraight, A. J.: *Our Dynamic World*, Chilton Books, New York, 1966.

COMPOSITION AND STRUCTURE OF THE ATMOSPHERE; TEMPERATURE PHENOMENA

CHAPTER 2

The atmosphere is an ocean of air surrounding the earth to depths of thousands of miles. It is an integral part of the planet, rotating and revolving around the sun together with the solid earth. Although invisible, it is essential to life, and its variations in temperature and moisture from one part of the earth to another greatly influence human activities.

COMPOSITION OF THE ATMOSPHERE

The atmosphere is made up of a relatively stable mixture of gases. The proportions of the gases not including water vapor are nearly uniform over the earth's surface and vary little up to elevations of over 100 miles. Two gases, nitrogen and oxygen, constitute over 99 per cent of air after water vapor and dust particles have been removed (Table 2–1).

In nature, the air is never completely devoid of moisture, nor is it free of solid impurities or dust particles. Water vapor is a variable ranging, close to the surface, from as little as 0.02 per cent of the air in some desert environments to over 4 per cent in the humid tropics. Although a minor component in terms of volume, it plays the most important role in influencing atmospheric conditions. It affects weather and climate in these respects:

1. It is the source of all condensation (conversion of vapor to a liquid or solid) and precipitation. Water vapor is the only gas in the atmosphere that condenses under normal atmospheric temperatures. Furthermore, the amount of moisture in the air directly determines the type and amount of precipitation (rain, snow, sleet, hail) that occurs.
2. It is the most important gas in the atmosphere that absorbs both solar radiation or energy and terrestrial (earth) radiation or energy. Thus it plays a major role in regulating the temperature of the air.
3. It is the source of latent heat, or stored-up energy. When liquid or ice is converted to water vapor, heat energy is expended. For example, it takes 79 calories of heat to convert 1 gram of ice to water and 607 calories to convert 1 gram of water to vapor. The heat utilized in the evaporation process remains stored in the water vapor as potential energy. The stored-up energy, or latent heat of vaporization, is released in the atmosphere when water vapor is converted or condensed back again into liquid or ice. This latent heat of condensation is a principal source of energy for violent storms such as hurricanes, tornadoes, and thunderstorms.
4. Water vapor affects the rate of cooling of the human body and thus influences *sensible temperature*. Sensible temperature is the tempera-

TABLE 2–1 COMPOSITION OF DRY AIR BY VOLUME

GAS	PER CENT
Nitrogen (N_2)	78.08
Oxygen (O_2)	20.95
Argon (A)	0.93
Carbon dioxide (CO_2)	0.03
Neon (Ne)	Trace
Helium (He)	
Methane (CH_4)	
Krypton (Kr)	
Nitrous oxide (N_2O)	
Hydrogen (H_2)	
Xenon (Xe)	
Ozone (O_3)	
Sulphur dioxide (SO_2)	
Nitrogen dioxide (NO_2)	
Iodine (I_2)	
Ammonia (NH_3)	
Carbon monoxide (CO)	

ture that the body feels; it often deviates from the actual temperature recorded by a thermometer. The human body generates heat, and the amount of moisture in the air affects the rate at which this heat is lost from the body. High humidities on a warm day retard loss of heat and cause the sensible temperature to exceed the actual temperature. High humidities on a cold day, on the other hand, make the body feel colder than it really is, if the skin becomes moist and evaporation, which is a cooling process, then occurs.

The atmosphere is never completely free of dust particles; there may be as many as 100,000 particles per cubic centimeter in the lower atmosphere over urban and industrial areas. Dust particles influence weather conditions in the following ways:

1. They reflect and scatter radiation from the sun. We may look upon them as tiny mirrors reflecting short waves of solar radiation and thus causing such phenomena as the blueness of the sky, the varied colors of sunsets and sunrises, and the occurrence of twilight and dawn. An unusual amount of dust in the atmosphere, as during periods of widespread volcanic activity, may appreciably reduce the temperature at the surface of the earth. Over large cities, smoke and dust act as a screen that reduces incoming sunlight, and as a blanket that retards outgoing terrestrial radiation, with consequent effects on local temperature conditions.
2. Certain types of dust particles, such as salt spray, coal and oil smoke, and sulphur trioxide, act as nuclei on which condensation of water vapor begins. Dust particles of this type are known as *hygroscopic nuclei* and are numerous over cities and coastal areas. In such localities there is considerably more cloudiness and fog than in areas that have fewer hygroscopic nuclei.

Carbon dioxide, oxygen, and ozone (triatomic oxygen) also influence weather conditions, though to a lesser degree. They absorb small amounts of solar energy and earth radiation and thus act as regulators of temperature at the surface of the earth. Although the average amount of carbon dioxide in the air is small, it varies considerably from one locality to another. Appreciable fluctuations have been observed over large cities and over areas of volcanic activity, with resulting effects on local temperatures. It is likely that industrial combustion associated with modern technological society is increasing the total amount of carbon dioxide in the air. The result may be a slight warming of the earth's atmosphere over a period of time, as the increased carbon dioxide acts to retard terrestrial radiation that would otherwise be released. In addition to the effects already mentioned, ozone acts as a filter screening out ultraviolet rays in upper air.

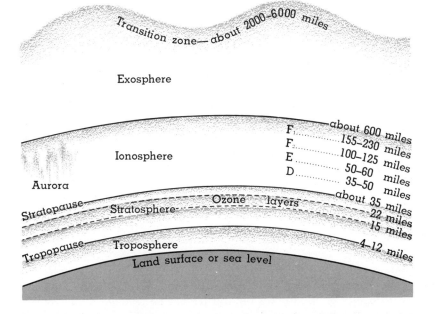

Transition zone—about 2000–6000 miles

Exosphere

Ionosphere

F .... about 600 miles
F₂ .... 155–230 miles
E .... 100–125 miles
D .... 50–60 miles
.... 35–50 miles

Aurora

Stratopause — Stratosphere — Ozone layers — about 35 miles
— 22 miles
— 15 miles

Tropopause — Troposphere — 4–12 miles
Land surface or sea level

FIG. 2-1. Generalized diagram of the structure of the atmosphere. **D** layer is slightly ionized and reflects radio waves well below the range of broadcasting frequencies. **E** layer reflects radio waves of 500–1500 kilocycles, the **AM** band. **F** layer (**F₁** and **F₂**) reflects waves of 1500 kilocycles to 15 megacycles; this range includes the **FM** band. **D** and **E** layers exist only in the daytime; **F₁** and **F₂** merge into one at night.

The earth's atmosphere is considerably different in composition from space, or from the atmosphere of other planets. Space is made up of about 90 per cent hydrogen and 9 per cent helium in volume, compared with about 0.00005 per cent and 0.0005 per cent of these gases, respectively, in the earth's atmosphere. Both gases are very light and have probably drifted away from the atmosphere of the earth over a long period of time. The earth's atmosphere has a much higher percentage of water vapor and free oxygen than the atmosphere of any of the other planets.

STRUCTURE OF THE ATMOSPHERE

The atmosphere extends to a height of several hundreds of miles. The air is compressible and is a great deal more dense near the surface of the earth, so that over one-half the mass of the atmosphere is concentrated in the lower 18,000 feet. At 18,000 feet the air is so rarefied that supplementary oxygen must be provided for air travelers and mountain climbers if they are not to suffer discomfort and a hazard to health.

The atmosphere may be thought of as being divided into a number of roughly concentric spherical shells, or strata. These may be defined and classified in a variety of ways, some more detailed than others. But for our purposes it will suffice to consider the various strata as being combined into four major layers (Fig. 2–1): (1) troposphere, (2) stratosphere, (3) ionosphere, and (4) exosphere.

The troposphere is the lower layer of the atmosphere, reaching to a height of 12 miles over the equator and gradually decreasing toward the poles to about 4 miles. Troposphere means "region of mixing"; the layer is so called because it is characterized by vigorous movements of ascending and descending currents. The ascending currents are, for the most part, convectional updrafts, originating where the earth's surface is heated. The troposphere varies in depth with the seasons—it is deeper in the summer and shallower in the winter. Most of the water vapor and a great many of the dust particles in the atmosphere are found here. The presence of water vapor and dust, associated with vertical motions of air, results in clouds, storms, and various weather changes. Above the troposphere storms and weather changes do not occur, although turbulence may exist in the lower stratosphere. Generally, the air becomes colder aloft in the lower atmosphere. The decrease in temperature with increasing elevation, which averages about 3.5° F. per 1000 feet, is known as the *lapse rate* and exists only in the troposphere. Above the troposphere the lapse rate is zero or negative. Finally, the troposphere is characterized by the great density of its air. By far the largest percentage of the mass of the total atmosphere is found in the troposphere, and the air in the upper atmosphere is exceedingly thin and rarefied.

Above the troposphere there is a layer of air, extending to elevations of about 35 miles, known as the stratosphere. The boundary between the troposphere and stratosphere is called *tropopause*. The stratosphere is characterized by the absence of storms or changing weather conditions, and by a negative lapse rate. At elevations of 15 to 22 miles, there are layers of ozone (triatomic oxygen) which absorb ultraviolet waves and become superheated. Temperatures of 30 to 40° F. have been measured in these ozone layers at times when temperatures of −60 to −100° F. have obtained in other parts of the stratosphere, both above and below. The ozone absorbs most of the sun's ultraviolet radiation, allowing only a small amount to reach the surface of the earth. Without this filtering effect, ultraviolet radiation would reach the surface of the earth in excessive amounts, harmful to life.

The ionosphere extends from the top of the stratosphere to heights of about 600 miles. It contains layers of ionized air that reflect radio waves, including those which fall within the range of broadcasting frequencies. The lower ionized layers (D and E) exist only in the daytime, but the higher layers (F_1 and F_2) merge into one that is present at night. Broadcasting radio waves are reflected back from this higher layer, and reception of far distant stations may thus often be achieved during the night hours.

The aurora borealis and the aurora australis (Northern and Southern Lights) are among nature's most spectacular displays. They result from the bombardment of showers of highly charged particles thrown from the sun. The auroras are found at about the same elevations as the radio-reflecting layers. The shape and form of the auroras vary greatly and depend upon the interaction of the charged particles and the earth's magnetic field.

The exosphere extends above the ionosphere to heights of perhaps 6000 miles. It is the transition zone between the earth's atmosphere and space. Although the air is exceedingly rare here—thinner than in a man-made vacuum tube—traces of the earth's atmosphere can be detected. The boundary between space and the atmosphere is not a precise line but a relatively undefined layer several hundred miles in thickness. As elevations in the exosphere increase, the frequency of molecules becomes less. At great heights, the density is so low that collisions between particles rarely occur. The exact height of the top of the exosphere is not certain. Recently, several belts of intense radiation 1000 to 3000 miles above the surface of the earth were discovered. They are known as Van Allen belts, after the scientist who discovered them during the International Geophysical Year (1957-58). It is thought that the radiation particles comprising these belts may be produced by the interaction of the sun with the earth's geo-magnetic field. Though they appear to be divorced from the normal atmosphere, their fluctuating intensity may have some relationship to the earth's weather cycles.

Although there are direct and indirect relationships between upper air and the weather experienced on earth, the actual day-to-day changes in weather occur only in the troposphere. The discussions to follow are concerned with the elements and controls of weather that exist in the troposphere and at the surface of the earth.

TEMPERATURE PHENOMENA

WHAT MAKES THE WEATHER

The condition of the weather at any time and place results from the interrelationship of three basic elements:

Temperature
Moisture
Air pressure and winds

These elements may be drastically altered from time to time and from place to place. The variations occur as the result of the following controlling factors.

Angle of the sun's ray and duration of day
Distribution of land masses and water bodies
Air masses and fronts
Altitude
Topography
Semipermanent low- and high-pressure centers
Ocean currents
Storms

The three basic elements and their controls occur in a complex inter-action that produces the weather at any given place and time. It would, of course, be impossible to understand them if we attempted to study them all together, so we shall have to separate them, somewhat artificially. For the balance of this chapter we shall discuss the heating processes of the earth and the controls that cause their variation.

HEATING PROCESSES

Solar radiation, or insolation, is the ultimate source of atmospheric heat, or energy. Its effect is not direct, however. Think of the fact that, as we climb a mountain or ascend in an airplane from the earth toward the sun, temperatures become steadily lower, rather than higher, as one might expect. The mechanics of heating the atmosphere are thus not simple, and they are decisively influenced by phenomena at the surface of the earth. The four heating processes that are *directly* responsible for warming the atmosphere are these:

Conduction
Absorption of solar radiation
Absorption of terrestrial radiation
Latent heat of condensation

Conduction. Conduction is the process by which heat is transferred through matter without any transfer of the matter itself; for example, when one end of an aluminum rod is heated, the other end soon becomes warm. When one end of a wooden rod is heated, however, the other end remains cool. The reason is that aluminum is a good conductor of heat and wood a poor one. When two objects are in contact, heat may pass from one object to the other by conduction. Thus during the day the earth's surface may be warmed by conduction. Conversely, at night the earth's surface may become colder than the air and the air may conduct some of its heat to the ground. In this way, the air tends to come to the same temperature as the surface of the earth with which it is in contact. Conduction is a minor heating process, however, because air is a poor conductor and the process is slow, consequently affecting only the lowest

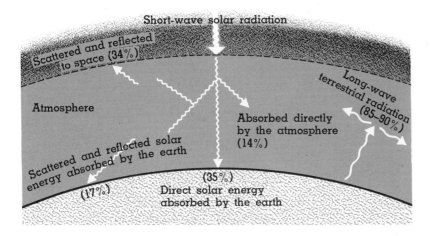

FIG. 2-2. Distribution of solar and terrestrial radiation. Much of solar energy is in the form of short waves, and only about 14 per cent is absorbed directly by the atmosphere, as an average. About 52 per cent of the solar radiation reaches the surface either directly or scattered and reflected by the atmosphere. On the other hand, 85 to 90 per cent of the long-wave terrestrial radiation is absorbed directly by the atmosphere.

2 or 3 feet of air. Heating of the air by conduction is largely restricted to the daytime and the summer season.

Absorption of solar radiation. Absorption of solar energy normally takes place largely so far above the surface of the earth that it, too, is a minor heating process. Consider the fact that, on a clear day in winter, the temperature near the surface remains very low in spite of a bright sun. The reason is that solar energy takes the form of short wavelengths, and the earth's atmosphere can absorb directly only a relatively small amount. Water vapor is the atmospheric gas primarily involved in absorbing solar radiation, though oxygen and carbon dioxide perform this function to a lesser degree. Hence, the more moisture in the air, the more solar energy absorbed in the atmosphere and the less reaching the surface. On the average, about 14 per cent of solar energy is absorbed by the atmosphere. Several times as much passes through the atmosphere to heat the earth's surface (Fig. 2–2).

Absorption of terrestrial radiation. The most important process in the heating of the atmosphere is the absorption of radiation from the earth. That is why there is a *decrease* in temperature as a balloon ascends and gets farther away from the surface of the earth.

Any land surface that is not covered with ice or snow readily absorbs the sun's short-wave energy. Some types of land surfaces in particular, such as those covered with forests, or plowed fields of black soil, absorb much and reflect very little of the solar radiation that reaches them. Water absorbs about the same amount of solar energy as most land surfaces. The earth's surface is thus heated by absorption of solar radiation and, in turn, becomes a radiating body, giving off heat in the form of long wavelengths. It is thus important to remember that solar radiation is in the form of *short waves* and terrestrial radiation in the form of *long waves*. Water vapor, oxygen, and carbon dioxide in the atmosphere can absorb up to 90 per cent of the earth's long-wave radiation and only about 14 per cent of the sun's short-wave radiation. This explains the so-called "greenhouse effect": the atmosphere acts like a pane of glass, allowing much of the short-wave solar radiation to pass through but greatly retarding the passage of the earth's long-wave radiation (Fig. 2–2).

As noted earlier, water vapor is the gas that is most active in absorbing both types of radiation. This may be demonstrated by comparing a desert and a humid area. In a desert region having a small amount of water vapor in the air, little solar radiation is absorbed in the atmosphere and daytime temperatures at the surface are high; at night, on the other hand, the dry air permits a rapid escape of the earth's radiation, with consequent low temperatures. In the Sahara, daytime temperatures of over 90° F. have been observed, followed by night temperatures below freezing. In a humid area, the daytime temperatures at the surface are lower than in the desert and nighttime temperatures much higher, with consequent relatively small temperature differences between day and night. In humid coastal areas, the daily range of temperature (difference between maximum and minimum temperatures in a 24-hour period) is usually 15–25° F. In general, the amount of moisture in the air is the most important control of the daily range of temperature in a given locality.

Radiation from the earth is a continuous process. During the daylight hours up to midafternoon, more solar energy is usually received than is radiated by the earth, resulting in rising temperatures. At night, however, solar energy is not received at the surface, and a continual loss of heat by earth radiation results in a drop in temperature. Cooling at night is even more pronounced if snow covers the ground, since snow is a poor conductor and allows little heat from the ground to escape to the surface.

Latent heat of condensation. A significant amount of the solar energy that reaches the earth's surface is consumed in evaporation (conversion of water to vapor). This energy is contained in the water vapor in a latent form. When the vapor condenses (converts to liquid form), this latent or potential energy is released to the atmosphere and is a major heat-adding process.

TRANSFERRING PROCESSES

The heat acquired by the atmosphere by the absorption of solar and terrestrial radiation, conduction, and latent heat of condensation is distributed from one part of the atmosphere to another by two methods. Horizontal transfer of heat is accomplished by *advection*, or the horizontal movement of air. Transfer by winds is the most important means of distributing heat from overheated to heat-deficient areas over the earth as a whole, and it therefore accounts for many of the day-to-day weather changes. A southerly wind in the United States Midwest results in very mild winter temperatures. Similarly, northerly winds from higher latitudes may produce a decrease in temperature of many degrees within a short period of time.

Vertical transfer of heat is accomplished by *convectional currents*. Surface air expands when heated and is forced to rise by the surrounding cooler and denser air. Convection is the most important means of transferring heat acquired by surface air to higher layers of the atmosphere.

Latent heat of condensation may be looked upon as a transferring as well as heat-adding process. Heat consumed at the surface by evaporation is carried by water vapor as it moves from the surface, vertically and horizontally, to various areas of the atmosphere. When condensation occurs, the latent heat is released.

The average temperature of the earth as a whole has remained about the same over several centuries. The incoming solar energy has been equal to the heat lost to space by terrestrial radiation. This balance, however, does not necessarily exist at specific localities. At tropical latitudes, incoming solar energy exceeds out-going radiation, and at high latitudes the reverse is true; but there is a continuous transfer of heat over the earth as a whole by winds and ocean currents.

CONTROLS OF TEMPERATURE

Temperature conditions at the surface of the earth vary greatly with place and time. These variations are created by the following controls: (1) the angle of the sun's ray, (2) length of day, (3) nature of the surface, (4) atmospheric conditions, (5) fluctuations in solar radiation, and (6) variations in the distance from the earth to the sun. The first three are of major significance, the others relatively minor.

Angle of the sun's ray. The sun is the original source of all heat in the earth's atmosphere. Thus the angle at which the sun's ray strikes the surface is the most significant control of temperature for the earth as a whole. A vertical ray delivers considerably more energy at the surface than an oblique ray, for two reasons. First, the energy delivered by the vertical ray is concentrated in the smallest possible area. Second, a vertical ray passes through a minimum thickness of atmosphere; an oblique ray, passing through a thicker segment, is weakened by the greater amount of reflection and absorption (Fig. 2–3). The weakening effect of the at-

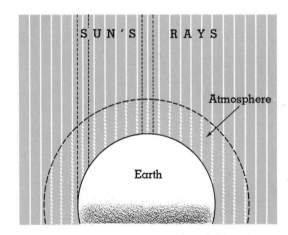

FIG. 2-3. The vertical or perpendicular ray of the sun delivers the most energy to the surface of the earth because it is concentrated over the smallest area of the surface and passes through the minimum amount of atmosphere, which absorbs, scatters, and reflects much of the solar radiation.

mosphere may be readily recognized by looking at the sun at noon, and again at sunset, on a bright day. It is not possible to look directly into the sun at noon, whereas it is quite comfortable to do so at sunset. Since the distance between sun and earth is the same, the major difference is that the slanting ray at sunset must pass through a greater amount of atmosphere.

The latitude of a given locality determines the angle of the sun. In tropical latitudes, the sun is high in all seasons and cold winters are therefore unknown. In polar latitudes, the sun's ray is always slanting and summers are cool. In midlatitudes, the angle of the sun is relatively high during one part of the year and low during another; cold winters and warm summers are the result. At latitude 45° north, for example, the noonday sun is 68½° above the southern horizon during the summer solstice and only 21½° above the southern horizon at the time of the winter solstice.

Length of day. The latitude of a given locality also determines the length of time the sun is above the horizon. At the equator, the daily daylight period is 12 hours throughout the year. At all other localities, the length of day varies with the season. For example, at latitude 45° north the sun is above the horizon 15 hours and 37 minutes during the summer solstice, the longest day of the year; and only 8 hours and 46 minutes during the winter solstice, the shortest day of the year. At the Arctic and Antarctic Circles the sun is above the horizon 24 hours during one solstice and below the horizon 24 hours at the time of the other solstice, 6 months later.

In general, the amount of solar energy received at a given location depends on the angle of the sun and the length of day. These are the most important temperature controls over the earth as a whole and largely explain the variations in temperature from place to place and from one season to another.

Nature of the surface. Land surfaces warm more rapidly and to a higher degree than water bodies when exposed to the same amount of solar energy; land surfaces also cool faster and to a lower degree. Since water is somewhat transparent, solar energy may penetrate it to depths of up to several hundred feet. Solar energy in water is therefore spread over a much greater volume than on land, where it is concentrated at the surface. About 90 per cent of the solar energy that reaches water spreads over the upper 30 feet, and small amounts may penetrate much deeper. Water also tends to be in constant motion due to convectional systems, waves, currents, and tides, and this disseminates the solar energy over large areas. On land, no horizontal or vertical mixing can take place.

The amount of heat utilized in the evaporation process is also a factor in the differential rates of heating of land and water. On land, actual evaporation is usually considerably less than possible or potential evaporation, whereas over water surfaces, evaporation is always at the maximum. It is probable that about one-third of the solar energy reaching the surface over the oceans is consumed in the evaporation process. Land surfaces do, however, differ from each other in the amount of evaporation that takes place—with a maximum in humid tropical areas and a minimum in the interior locations of continents at midlatitudes. Finally, the specific heat of water is about two and one-half times that of soil, which means that it requires about two and one-half times more heat to raise the temperature of a volume of water than an equal volume of soil.

For the earth as a whole, land surfaces reflect about the same amount of solar radiation as water. There is, however, a great variation among different types of land surfaces. Ice or snow-covered surfaces reflect considerably more solar radiation than water—perhaps as much as 90 per cent of the total solar energy received. On the other hand, black, exposed soil, forest, and grass-covered surfaces reflect relatively small amounts of solar energy and hence absorb a higher percentage of solar energy than water bodies.

For the reasons enumerated, locations in the interior of midlatitude continents experience higher temperatures than coastal locations during the day and in the summer season, and lower temperatures at night and in the winter season. Consider, for example, Seattle, Washington, close to the coast, and Moorhead, Minnesota, in the interior of the continent, both at about the same latitude: In Seattle, the coldest month averages 40° F.; in Moorhead, 4° F. On the other hand, the warmest month averages 64° F. in Seattle and 70° F. in Moorhead. On the average, there is only one day a year with temperatures below 32° F. in Seattle but 174 days in Moorhead. Seattle experiences only one day a year with temperatures above 90° F. whereas Moorhead experiences 9 days of 90° F. or more.

Atmospheric conditions. Although they constitute a minor factor in explaining temperature differences for the earth as a whole, atmospheric conditions over a local area do influence the amount of solar energy that is reflected or absorbed on its way to the surface of the earth. The amount of water vapor and dust in the air determine to a large degree how much energy reaches the surface. Since the dust and moisture content of the lower atmosphere inevitably vary from place to place, this factor has an important bearing on local temperature.

Fluctuations in solar radiation. The energy received from the sun fluctuates up to 3 per cent from the average of about 1.94 calories per square centimeter per minute. These variations take place over a period of a few years and are associated with sun-spot activity. It is not known what, if any, direct effect on temperature conditions at the surface of the earth these fluctuations in the solar constant produce.

Variations in the distance from the earth to the sun. The average distance from the earth to the sun is 93 million miles, but the earth travels in an elliptical orbit and is thus about 3 million miles closer to the sun in January, at the perihelion, than in July, at aphelion. This variation is a minor factor in controlling temperature, since there is a severe cold season in the Northern Hemisphere at the time the earth is nearest the sun. In fact, the lowest temperatures recorded on earth in midlatitudes occur at this time. The angle of the sun's ray, length of day, and distribution of land and water are of far greater importance.

VERTICAL DISTRIBUTION OF TEMPERATURE

Under normal conditions, there is a decrease of temperature aloft. The rate of decrease varies with the locality, the time of day, and the season, but it averages approximately 3.5° F. per 1000 feet. As noted earlier, this average rate of decrease in temperature with increasing elevation, or lapse rate, exists to the heights of the tropopause. The fact that temperatures are highest at low elevations and become colder aloft illustrates the fact that it is the surface of the earth, rather than the sun, that is the major direct source of heat for the atmosphere.

Temperature inversions. Within the first few hundred feet of the atmosphere, temperatures may at times increase with elevation. Such occurrences are called *temperature inversions*; they may be either static or dynamic. Static temperature inversions occur when the earth's surface becomes colder than the air above, as a result of loss of heat by radiation, and in turn chills the lower layer of air. Inversions of this type are not uncommon at night in midlatitudes during the winter season. Ideal conditions for their development are (1) a clear sky, permitting rapid loss of heat by terrestrial radiation; (2) cold air with little water vapor, so that a minimum of absorption of terrestrial radiation takes place; (3) a long night, providing an extended period of time when terrestrial radiation may take place without incoming solar radiation; (4) relatively quiet air, so that little mixing takes place and the surface air becomes cold by radiation and conduction; and (5) a snow-covered surface, which reflects solar energy during the day and retards the loss of heat from the ground at night.

A dynamic temperature inversion results from an advective movement of air, where cold air blows in, displacing warmer air and wedging it aloft. The formation of smog in Los Angeles and other cities is due to a large degree to this type of temperature inversion (Fig. 2–4). Once the cold air

is blown in, wedging the warm air above, smoke and exhaust fumes are carried upward by air currents through the cool air to the base of the warm-air layer. The air currents cannot readily penetrate the warm air and spread out horizontally. The warm air thus acts as a lid, holding the smoke, dust, and exhaust fumes close to the surface of the ground.

Inversions also take place in the upper atmosphere. They are most commonly associated with semipermanent high-air-pressure areas and are the result of the settling and warming of large masses of air. Upper-air inversions usually restrict the upward movement of air currents and consequently are often associated with dry climates.

Static temperature inversions become exceptionally well developed in areas of irregular topography. The colder, denser air slips downslope into valleys, where it accumulates in pools. The movement of cold air downslope at night is called air drainage and causes markedly lower temperatures in lowlands than in neighboring uplands. This phenomenon must be taken into account in planting commercial crops that are sensitive to frost.

Frost is said to occur when the temperature has fallen to 32° F. or below. It is most commonly associated with static and dynamic temperature

FIG. 2-4. Cleveland, Ohio, under smog. Smog development over an industrial city is the result of restriction by a temperature inversion of the vertical movement of dust and air pollutants. (U.S. Public Health Service from The Cleveland Press.)

inversions and usually occurs first in lowland areas. The threat it presents to agricultural production can be minimized by proper selection of the sites at which frost-sensitive crops are planted. Coastal areas along large bodies of water are particularly suitable for this purpose, since water cools slowly and thus tends to keep the bordering lands at higher temperature than inland areas. In the fruit belts close to the largest lakes in Michigan, Wisconsin, Ohio, and Ontario, sensitive fruit crops are grown successfully at relatively high latitudes. Planting crops on the slopes and highlands, and avoiding lowlands subject to air drainage, can also minimize the hazard of killing frost.

Several other methods of frost protection are current in various parts of the country. In the citrus groves of California and Florida, heaters are placed among the trees, often as many as sixty per acre (Fig. 2–5). They actually add heat to the lower air and, under ideal conditions, may keep the temperature as much as 12° F. warmer than in adjacent areas. In addition to giving off heat, they create small convectional currents that tend to mix the lower air and prevent the development of a cold layer at the surface. Wind machines are also used in citrus groves. Although these machines do not add heat, they too mix the lower air, preventing a layer of air from remaining stationary at the surface and becoming chilled by the ground.

There are several frost-protection measures based on retarding the loss of heat by terrestrial radiation. The simplest means is to cover sensitive plants with some nonmetallic material such as paper, straw, or cloth. The

FIG. 2-5. Glowing heat bricks in a citrus orchard, one device to prevent frost damage. Temperatures may be kept elevated several degrees by means of such properly located heating elements. (Mobil Oil Corporation, New York.)

FIG. 2-6. Weather-recording instruments. Instrument shelter on the left houses temperature-recording instruments. This type of shelter, in standard use, has a sloping double roof, and all sides are louvered to permit free movement of air while protecting the instrument from sun, rain, and snow. The instruments on the left are the standard 8-inch rain- and snow-weighing gauges. Evaporation pan in center of picture records evaporation data. (Environmental Science Services Administration, U.S. Department of Commerce.)

purpose of the cover is not to keep out the cold but to keep in the heat that would otherwise be radiated from the ground. Smudge pots give off hygroscopic smoke particles, which create a smoglike blanket over citrus orchards, thus retarding cooling by radiation. Smudging has proven not to be highly effective, however. Also, the smoke is a public nuisance, and when it remains in the air during the day, it delays surface warming. A frost-prevention measure utilizing the principle that water is a conserver of heat is widely employed in the cranberry-producing areas of Wisconsin and New England, where the bogs in which the cranberries are grown are flooded when frost seems probable. Since the water surfaces cool less than the land, the hazard of a killing frost is minimized.

OTHER PERTINENT TEMPERATURE DATA

All statistical temperature data are based on the accumulation of one central fact—the temperature of the air at a given place and time. Accurate, standardized observations and records are essential if temperature information is to be meaningful and subject to comparison from one place to another and one time to another. Such observations have been made for over a century in Europe and for 95 years in the United States. To obtain a standardized reading, the thermometer should be located in an instrument shelter that provides shade but allows exposure to freely moving air; the shelter should be about 4 feet above the ground, over sod if possible, and away from buildings, roads, and other structures (Fig. 2–6). An accurate thermometer properly placed will record temperatures markedly different from those registered by an instrument that is not precise in its readings or is placed incorrectly.

Daily temperature data. Several kinds of information concerning temperatures over the course of the day are significant. The maximum and minimum temperatures over a 24-hour period are essential, as they are the basis for both the daily mean and the daily range. The daily mean is calculated by adding the maximum and minimum together and dividing by two. This statistic is widely used to compare temperatures from place to place and from time to time. The daily or diurnal range is the difference between the maximum and minimum temperatures during a 24-hour period. In desert areas, where there is little moisture in the air, the daily range may be several times what it is in highly humid areas.

The *thermograph* is an instrument that makes a continuous record of temperature, from which the daily march of temperature can be determined. This indicates the rise and fall of temperature over a 24-hour period, reflecting the balance between incoming solar radiation and outgoing terrestrial radiation. Normally, the lowest temperature is recorded at dawn and the highest between 2:00 P.M. and 4:00 P.M. The 2- to 4-hour lag between the period of maximum solar energy—usually at noon suntime—and the period of maximum temperature may be explained by the "greenhouse effect," discussed earlier. Recall that the earth's long-wave radiation is more important in warming the atmosphere than the sun's short-wave radiation; and it takes the earth 2 or 3 hours to reach its maximum temperature after the sun has reached its zenith.

Sensible temperature. The actual temperature and the sensible temperature—what the human body feels—may deviate considerably. Humidity, wind, and sun absorption may retard or accelerate heat loss from the body and causes the sensible temperature to be higher or lower than the actual temperature. No instrument can accurately measure all factors influencing sensible temperature, but the wet-bulb thermometer (one whose bulb is covered with a saturated gauze or cloth) does measure one important factor—the moisture in the air; and this is a good indicator of sensible temperature.

Growing degree-days. Each species of plant has a basic temperature at which growth begins. For many commercial plants, this temperature is around 40° F. Furthermore, each crop requires a specific number of growing degree-days to develop from a seed to maturity. A growing degree-day is counted for each 1° F. of the daily mean temperature above the basic temperature. For example, let us assume the basic temperature for a particular species is 40° F., and about 1500 growing degree-days are needed for the plant to develop from seed to maturity. On a given day, if the daily mean is 55° F., 15 growing degree-days are accumulated; if on the following day the daily mean were to be 72° F., 32 growing degree-days would be added. The plant will be ready for harvest whenever the total of 1500 growing degree-days is accumulated—whether it is quickly, over the course of a few weeks, or more slowly, over several months.

Frost-free season. The number of consecutive days in which the temperature remains above 32° F. is known as the frost-free or growing season. It varies from 365 days in some tropical areas to none in certain polar or highland regions. The frost-free season is highly important, as it limits the kinds of crops that can be grown in any specific locality. Frost-free seasons of 365 days permit the production of most tropical crops, such as rubber, cacao, and bananas. Growing seasons of 200 or more days permit the cultivation of subtropical crops—cotton, rice, and tea, for example. On the other hand, a short growing season—120 days or less—restricts production to a few types of hardy grains and root and hay crops; and a frost-free season of less than 90 days virtually eliminates commercial agriculture. But the average length of the frost-free season may not in itself be sufficient information for detailed agricultural planning if departures from the average are large and frequent. A better statistic for this purpose is the *effective frost-free season*, or the frost-free period in 4 out of 5 years. This may be anywhere from 10 to 50 days less than the average frost-free season, depending on locality; its use makes the risk of a killing frost very slight (Fig. 2–7).

Heating degree-days. A statistical device used to estimate fuel consumption is the heating degree-day. It is defined as a 1° F. departure of the daily mean below 65°. For example, if on a given day the daily mean

FIG. 2-7. Highly generalized map of the effective frost-free or effective crop season (length in days, 4 years out of 5) in the United States.

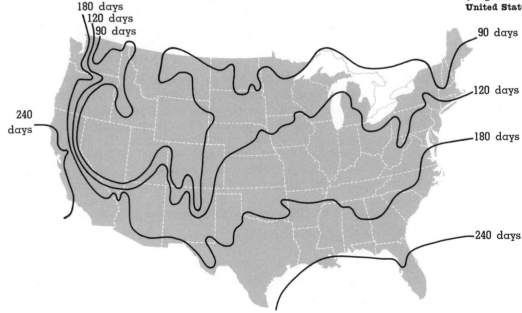

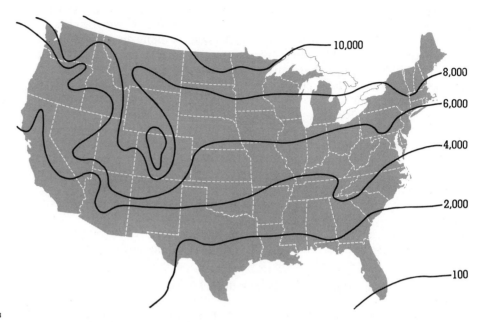

FIG. 2-8. **FIG. 2-8.** Generalized map of the average number of heating degree-days per year in the United States.

is 50° F., there are 15 heating degree-days; and if on the following day the daily mean were to be 20° F., there would be 45 heating degree-days and the fuel consumption would normally be three times as great as on the previous day. The number of heating degree-days in the United States per season varies from less than 100 in southern Florida to over 10,000 in northern Minnesota (Fig. 2–8).

Isotherms. Temperature distribution, whether over a limited area or over the entire earth, can best be shown by means of an isothermal map. An *isotherm* is a line drawn through points of the same sea-level temperature. Isotherms can be used to illustrate any statistical temperature data— daily means, annual temperatures, monthly means, etc. For the earth as a whole, January and July are the months of seasonal extremes of temperature. A map illustrating the behavior of isotherms during these months shows at a glance the pattern of seasonal temperature variations (Fig. 2–9).

A comparison of the January and July isotherms reveals the following:

1. There is a general decrease of temperature from the tropical latitudes to the poles.
2. Isotherms generally trend east-west, reflecting the importance of the angle of the sun's ray and the length of day as temperature controls.

3. There is a latitudinal shifting of temperature that follows the migration of the vertical ray of the sun.
4. The migration is greater over land masses than over water because of the differential heating and cooling of land and water.
5. The highest and lowest temperatures are found over land.
6. Isotherms are relatively regular in the Southern Hemisphere owing to the lack of large land masses at midlatitudes, whereas in the Northern Hemisphere they bend sharply equatorward in January and poleward in July over the continents.
7. Isotherms bend equatorward over cold ocean currents and poleward over warm ocean currents.
8. Isotherms are closer together over land in the winter season, reflecting steeper temperature gradients.

Annual range of temperature. The difference between the average temperatures of the warmest and coldest months is called the annual range of temperature. It varies considerably from one locality to another. It is a meaningful statistic, since it summarizes the seasonal extremes. Annual ranges of temperature are small in equatorial regions, where there is little variation in insolation over the year, and increase with increasing latitude. Ranges are also small over large bodies of water, which undergo little heating and cooling, and are greatest over continents at midlatitudes owing to the greater heating and cooling of land. Consequently, the annual range of temperature varies with the size of the land mass and the latitude—from almost zero at certain localities on the equator to over 110° F. in the interior of the Eurasian continent. The annual range of temperature for a given locality can be estimated from Fig. 2–9.

Annual march of temperature. A line drawn through the twelve monthly means gives the annual march of temperature. Usually, the months of January and July are the extremes. In the Northern Hemisphere, the lowest temperatures normally occur during the January 15–25 period and the highest during the July 15–25 period. The extremes of temperature usually lag about one month behind the migration of the vertical ray of the sun. Coastal or island locations, however, may have their lowest temperatures in February and the month of highest temperature may be delayed until August or September owing to the influence of large bodies of water. In southeastern Asia, where the monsoons are well developed, the warm month may be April or May, just preceding the coming of the summer monsoons. The summer monsoonal rains, bringing considerable cloudiness and rainfall, reduce the amount of solar radiation received at the surface.

Thermal anomalies. The average monthly temperatures for any given degree of latitude as they would exist on a homogeneous surface can be

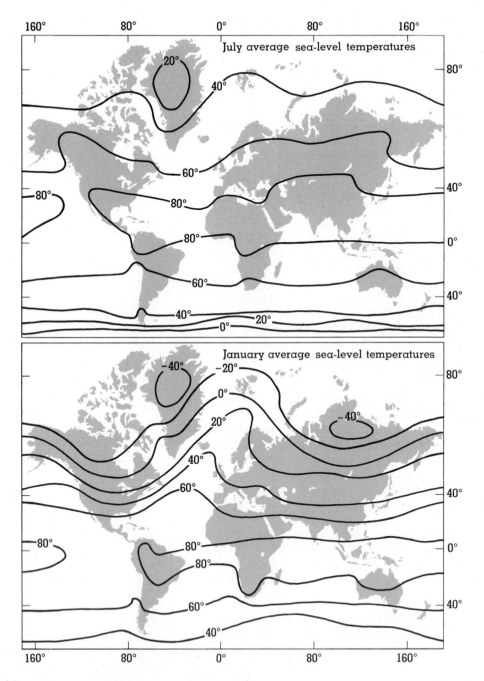

FIG. 2-9. Generalized maps of average sea-level temperatures.

determined by calculating the balance of incoming solar radiation and outgoing terrestrial radiation. The actual departure from such an average temperature for a given parallel is called a *thermal anomaly* and measures the influence of controls other than the sun, such as land and water distribution, ocean currents, prevailing winds, and air masses. The greatest anomalies are found in the Northern Hemisphere, where extensive land and ocean surfaces both exist. Because of an exceptionally warm ocean current, the coast of Norway has the largest positive winter thermal anomalies anywhere on earth. Some localities are over 45° F. warmer than the mean for their latitude in the month of January. The greatest negative winter thermal anomalies are in the interior of Siberia, where it is over 45° F. colder than the mean in January. These negative anomalies reflect the influence of a large land mass and extreme cooling.

New types of equipment developed recently now make it possible to gather temperature and other weather data in remote parts of the world for which detailed information has not heretofore been available (Fig. 2–10).

FIG. 2-10. An automatic atomic-powered unmanned weather station, Nomad I. These stations are designed to operate in remote areas unattended for periods of several years. The station radios wind speed and direction, air temperature, sea temperature, and barometric pressure. (National Bureau of Standards.)

SUGGESTED REFERENCES

Blair, T. A., and R. C. Fite: *Weather Elements*, Prentice-Hall, Inc., Englewood Cliffs, N.J., 1965.

Blumenstock, D. I.: *The Ocean of Air*, Rutgers University Press, New Brunswick, N.J., 1959.

Day, J. A.: *The Science of Weather*, Addison-Wesley Publishing Company, Inc., Reading, Mass., 1966.

Massey, H. S. W., and R. L. F. Boyd: *The Upper Atmosphere*, Hutchinson and Co. Ltd., London, 1958.

Molga, M.: *Agricultural Meteorology*, Office of Technical Services, U.S. Department of Commerce, Washington, D.C., 1962.

Riehl, H.: *Introduction to the Atmosphere*, McGraw-Hill Book Company, Inc., New York, 1965.

Sellers, W. D.: *Physical Climatology*, University of Chicago Press, Chicago, 1965.

Sullivan, W.: *Assault on the Unknown; The International Geophysical Year*, McGraw-Hill Book Company, Inc., New York, 1961.

Taylor, G. F.: *Elementary Meteorology*, Prentice-Hall, Inc., Englewood Cliffs, N.J., 1957.

AIR PRESSURE, WINDS, AND MOISTURE

3

AIR PRESSURE

Air pressure, the next of the weather elements we shall discuss, like the others is never constant for any length of time and varies considerably from one location to another. Unlike most atmospheric phenomena, however, pressure differentials at the same elevation are ordinarily not perceptible to the human senses, and changes in pressure are in themselves relatively unimportant. But differences in pressure do affect air motion, and this in turn may bring about rapid and noticeable changes in temperature and moisture. Thus air pressure and the resulting air movements are of the greatest importance as an *indirect* weather control.

The atmosphere is held to the solid and liquid earth by the force of gravity. It has a substance and weight that are measured in terms of air pressure. At sea level, this pressure averages 14.7 pounds per square inch. It fluctuates somewhat as changes occur in atmospheric conditions. The mercurial barometer is the most accurate instrument available to measure air pressure. It consists of a glass tube from which the air has been removed, with its open end immersed in a container of mercury (Fig. 3–1). Since the air exerts pressure on the mercury in the container, the height to which the mercury rises in the tube is a measure of the air pressure. When pressure is high, the column of mercury is higher than normal; when pressure decreases, the column drops. At sea level, the average height of the column of mercury is 29.92 inches, and this is considered the normal atmospheric pressure. Another unit of measurement in general use is the *bar*. Expressed in this unit, normal atmospheric pressure is 1013 millibars (= 29.92 inches).

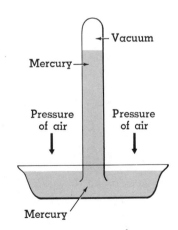

FIG. 3-1. Principle of the mercurial barometer.

FIG. 3-2. The barograph, an instrument used for the measurement of air pressure. The flexible metal chamber is partially exhausted of air, and springs inside the chamber are balanced against the outside air pressure. As the flexible chamber responds to changes in pressure, a permanent record is made by a pen on a slowly rotating drum. (Environmental Science Services Administration, U.S. Department of Commerce.)

TABLE 3-1 NORMAL AIR PRESSURES AT 1,000-FOOT INTERVALS

ELEVATION (feet)	AIR PRESSURE (inches)
18,000	14.94
17,000	15.56
16,000	16.21
15,000	16.88
14,000	17.57
13,000	18.29
12,000	19.03
11,000	19.79
10,000	20.58
9,000	21.38
8,000	22.22
7,000	23.09
6,000	23.98
5,000	24.89
4,000	25.84
3,000	26.81
2,000	27.82
1,000	28.86
sea level	29.92

The aneroid barograph is an instrument that measures air pressure and at the same time makes a continuous record of pressure change (Fig. 3-2). A flexible metal chamber is partially exhausted of air, and springs inside the chamber are balanced against the outside air pressure. The flexible chamber responds to changes in air pressure, which are recorded by a pen on a slowly rotating drum, thus creating a permanent record.

VERTICAL DISTRIBUTION OF AIR PRESSURE

The weight and density of air pressure decrease with elevation. At the height of 18,000 feet, air pressure is approximately one-half that at sea level (Table 3–1). The low pressure at elevations of 12,000 feet and above cause marked discomfort. Breathing is difficult because of the reduced amount of oxygen available for the blood to circulate to the tissues. The body adjusts by increasing the number of red blood corpuscles, thus making it possible for the blood to absorb the available oxygen more easily. In the South American highlands, Indian tribes that have lived at high altitudes for centuries have adapted to their environment by developing much greater lung capacity and larger hearts than people who inhabit lands of lower elevations.

HORIZONTAL DISTRIBUTION OF AIR PRESSURE

We do not as yet fully understand what causes the variations in air pressure from one location to another. We are sure, however, that at least two factors are involved: (1) temperature, and (2) the effects of deflection associated with the earth's rotation.

Air pressure varies with temperature; air expands when it becomes warmer and warm air consequently exerts less pressure than cold. Thus over the earth as a whole, regions of low temperature tend to have relatively high pressures and regions of high temperatures tend to be low-pressure areas.

Further, all moving bodies are deflected by the rotation of the earth—to the right in the Northern Hemisphere, to the left in the Southern Hemisphere. This deflection is called the Coriolis force, after the French scientist who first defined it mathematically in 1844. Bear in mind, however, that the Coriolis force is an *apparent* force; that is, the path of the moving body is curved with respect to the rotating surface of the earth, but not with respect to the stars. A rocket fired from the equator directly north toward the Arctic Circle will not continue to travel due north in relation to the surface of the earth because the earth's surface rotates faster at the equator than toward the poles. The rocket will drift toward the right, as observed from the earth (Fig. 3–3). Similarly, winds in the Northern Hemisphere, deflected to the right, will pile up air to the right, creating higher air pressure. In the Southern Hemisphere, winds deflected to the left develop higher air pressure to the left of horizontally flowing air.

The earth and its atmosphere receive heat from the sun and radiate approximately the same amount of heat into space. But this balance of heat input and output applies only to the earth as a whole and not to any specific area. Between about latitudes 35° south and 35° north, more heat is received from the sun than is radiated into space. On the other hand, from about latitude 35° poleward more heat is lost by radiation than is received from the sun. The surplus heat of the lower latitudes is transferred to the higher latitudes by atmospheric circulation and ocean currents.

In the surplus-heat areas of the tropical latitudes, warm air expands, rises, and flows toward the poles. This flow of upper air is deflected to the right in the Northern Hemisphere and to the left in the Southern Hemisphere, resulting in the development of the *subtropical jet stream* at about latitude 30° in both hemispheres. The subtropical jet stream is an upper wind that travels west to east around the globe at velocities of from 100 to over 300 miles per hour. The resulting damming up of air creates a high-air-pressure zone at latitudes 30° north and south, commonly referred to as the subtropical high.

Thus as a result of the latitudinal distribution of temperature and of the earth's rotation, there is a general zoning of sea-level pressures. At both poles there are high-pressure zones caused by the prevailing low temperatures; and at equatorial latitudes the year-round high temperatures create a low-pressure zone. These pressure belts, known as the polar highs

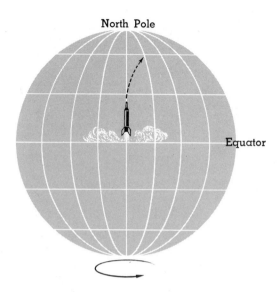

FIG. 3-3. Effect of the earth's rotation (Coriolis force).

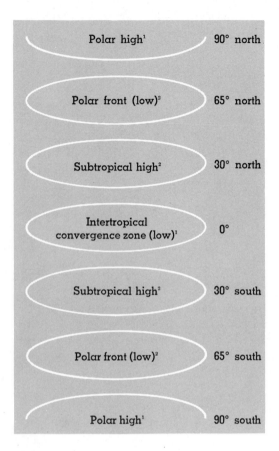

Polar high[1]	90° north
Polar front (low)[2]	65° north
Subtropical high[2]	30° north
Intertropical convergence zone (low)[1]	0°
Subtropical high[2]	30° south
Polar front (low)[2]	65° south
Polar high[1]	90° south

FIG. 3-4. Schematic diagram of the world's sea-level pressure cells (effects of unequal heating of land and water are neglected); 1: thermally controlled (due to differences in temperature); 2: dynamically controlled (due to the earth's rotation).

and the equatorial low, or intertropical convergence zone, are thermally controlled—that is, they are due to the unequal receipt of solar radiation.

At about latitude 30° in both hemispheres, there are irregular belts of high pressure, as previously mentioned, called the subtropical highs. These high-pressure belts result from the earth's rotation and the damming up of upper winds associated with the subtropical jet stream. Thus they owe their existence to *dynamic* rather than *thermal* factors. In the vicinity of the Arctic and Antarctic Circles (66½° N. and S.), the subpolar lows, or polar fronts, are low-pressure cells; they too are the result of dynamic rather than thermal conditions. These belts, or cells, are centers of pressure that are elongated in shape and trend in a general east-west direction (Fig. 3-4).

The heating and cooling of land also greatly modify the schematic worldwide patterns of air pressure previously discussed. Over land masses located in midlatitudes, steep high-pressure centers develop in the winter season as the result of low temperatures, and low-pressure cells appear in the summer (Fig. 3-5).

A comparison of the January and July isobaric* maps reveals these conditions:

1. Isobars generally trend east and west, reflecting the thermal and dynamic controls of pressure.
2. In general, the equatorial low, or intertropical convergence zone, subtropical highs, subpolar lows, and polar highs are well defined.
3. There is a seasonal latitudinal migration of the pressure systems that follows the vertical ray of the sun.
4. Isobars are more regular in the Southern Hemisphere, reflecting the lack of large land masses at midlatitudes.
5. There is a strong high-pressure cell in January over the Eurasian continent as a result of the very cold land mass.
6. The subpolar low in the Northern Hemisphere in January is divided into two well-developed low-pressure cells situated over Iceland and the Aleutian Islands.
7. During the summer, there is a well-developed thermally controlled low-pressure area over the southern portion of Asia produced by a merging of the equatorial low and the continental low-pressure areas.

WINDS AND ATMOSPHERIC CIRCULATION

Wind is air in horizontal motion. The motion results from differences in air pressure and always proceeds from high-pressure areas toward low-pressure areas. Winds in themselves may be

* An isobar is a line running through points of equal sea-level pressure.

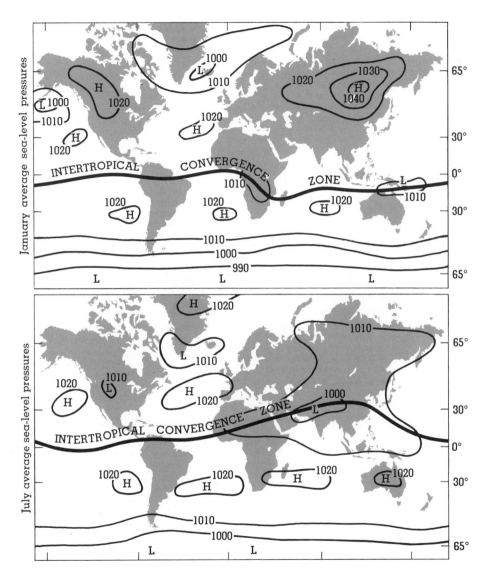

FIG. 3-5. Generalized maps of average sea-level pressures, in millibars (1 inch of air pressure = 1013.2 millibars). Each line is an isobar; that is, it runs through points of equal pressure at sea level.

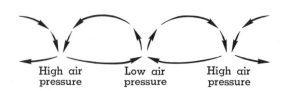

FIG. 3-6. **Highly generalized schematic diagram of the circulation of air in high- and low-pressure areas.**

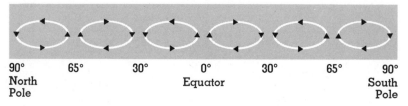

FIG. 3-7. **Schematic diagram of the circulation cells in each hemisphere.**

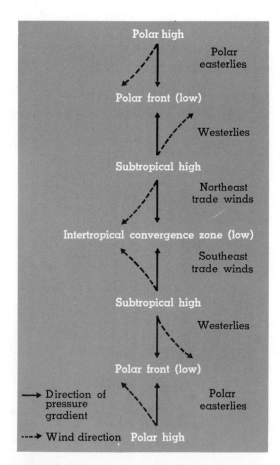

FIG. 3-8. **The effect of Coriolis force on the planetary winds.**

considered the most important single initiator of weather change, for clouds, storms, cold waves, and hot spells all depend upon winds for their horizontal movement.

Wind direction and speed are determined by the following controls:

1. The pressure gradient directly influences wind velocity—the steeper the gradient, or the greater the difference between the high- and low-pressure areas, the greater the speed of the wind.
2. Coriolis force, as previously explained, deflects the wind, as it does all moving bodies, to the right in the Northern Hemisphere and to the left in the Southern Hemisphere.
3. Centrifugal force affects the direction of the wind by pulling the air outward from its center of curvature.
4. Friction with the earth's surface decreases the speed of winds near the surface and influences their direction slightly.

Winds are always described in terms of direction and velocity. The velocity is measured in either miles or knots per hour, and it varies directly with the pressure gradient. Closely spaced isobars indicate steep gradients and strong winds; where isobars are far apart, the winds are weak. In speaking of direction, we name winds by the point of the compass *from* which they blow. Thus a horizontal motion of air from the northeast toward the southwest is a northeast wind. Wind blowing frequently from one direction is called a prevailing wind.

GENERAL CIRCULATION OF THE ATMOSPHERE

Throughout hundreds of years man has noticed that in some areas of the earth the winds blow predominantly from one direction all year; in other areas the prevailing direction changes with the seasons; and in still others the winds are so variable that there is simply no discernible

48

pattern. Despite these differences, the distribution of prevailing winds and pressures over the earth may be generalized into two major systems: (1) planetary circulation, and (2) monsoonal systems, best exemplified in southern Asia.

Planetary circulation. The distribution of air-pressure zones trending east-west gives rise to a generalized system of winds and pressure belts. The movement of air at the surface blows from high to low, and upper air blows generally in the opposite direction. Low-pressure areas are zones of converging winds and ascending currents, and high-pressure areas are zones of diverging winds and descending currents (Fig. 3–6).

The presence of the intertropical convergence zone (low), subtropical highs, polar fronts (low), and polar highs results in three generalized circulation cells in each hemisphere (Fig. 3–7):

1. Surface air blowing from the polar highs to the polar fronts (low) ascends and returns aloft to the poles.
2. Air moving from the subtropical highs to the polar fronts at the surface ascends and returns to the subtropical high as upper winds.
3. Winds blowing at the surface to the intertropical convergence zone rise and return aloft to the subtropical highs.

Thus the winds, like all moving bodies, are deflected to the right in the Northern Hemisphere and to the left in the Southern Hemisphere. The Coriolis force acts at right angles to the horizontal movement of the wind caused by differences in air pressure. For this reason, the planetary winds do not blow directly from high to low but are deflected in their course (Fig. 3–8). To estimate the deflection correctly, it is necessary to face in the direction in which the wind is blowing so that the drift of the deflection can be properly ascertained. Note that the velocity of the wind, which is the result of the pressure gradient, is not affected by the Coriolis force. Figure 3–9 shows instruments used to measure wind direction and velocity.

The planetary system of winds, then, results from the varying amounts of solar energy reaching the surface and creating thermally controlled pressure zones; and from the rotation of the earth, which causes dynamically controlled pressure zones; and it is modified by the Coriolis force (Fig. 3–10). The polar highs are areas of 30-to-30.5-inch pressures, within which the most conspicuous motion of air consists of descending currents. The polar easterlies are cold, dry winds blowing from the polar highs to the subpolar lows in the vicinity of the Arctic and Antarctic Circles. The subpolar lows are zones of stormy weather and variable winds, where the polar easterlies converge with the winds moving poleward from the subtropical highs. They are more strongly developed in the Southern

FIG. 3-9. Instruments used to measure wind direction and velocity. The wind vane on the left indicates wind direction, and an automatic record is obtained by an electrical circuit connecting the vane to a recording device actuated by clockwork. The three-cup anemometer, on the right, records wind speed. As the cups revolve, wind velocity is indicated on a dial. An automatic record of wind speed may be obtained on a meteorograph, along with wind direction. (Environmental Science Services Administration, U.S. Department of Commerce.)

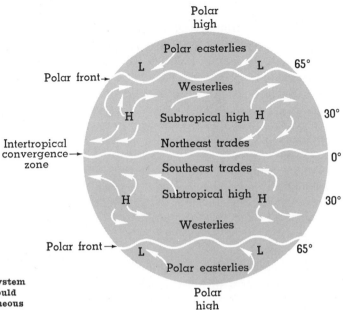

Polar
high

Polar easterlies

L   L   65°

Polar front→

Westerlies

H   Subtropical high   H   30°

Northeast trades

Intertropical
convergence →   0°
zone

Southeast trades

Subtropical high   H   30°

H

Westerlies

Polar front →   L   L   65°

Polar easterlies

Polar
high

FIG. 3-10. Model of the planetary system of wind and pressure belts as they would exist on a rotating earth and homogeneous surface (effects of unequal heating of land and water surfaces are neglected).

Hemisphere. The westerlies, between latitudes 30° and 65°, are humid winds, variable in a specific locality owing to the presence of moving cells of low and high pressure called cyclones and anticyclones (to be discussed in Chapter 4). The subtropical highs are areas of variable winds and calms with pressures averaging from 30.0 to 30.2 inches. They are the result of damming up of upper air deflected by the earth's rotation. The movement of air is predominantly in the form of descending currents, with consequent clear, dry weather conditions. The trade winds are consistent winds of 10 to 15 miles per hour and are usually characterized by fair weather. If they are forced to ascend over highlands, however, heavy rains may result. The equatorial low, or intertropical convergence zone, is an area of variable, feeble winds and ascending currents. It is a region of considerable cloudiness and heavy rainfall. The wind and pressure belts migrate seasonally, following the vertical ray of the sun, with migration greater over land.

The subtropical jet stream, found in the latitudes of the subtropical high in both hemispheres, is an integral part of the planetary system of winds, though it is an upper-air wind. It is a fast, narrow stream of air traveling west to east at velocities of from 100 to over 300 miles per hour, at elevations of 35,000–40,000 feet—about 2000–3000 feet below the tropopause—and at latitudes 30° to 35° in both hemispheres. It meanders widely, however, and may range as far equatorward as latitude 20° and to latitude 45° at the poleward extremes (Fig. 3–11). It is 25 to 100 miles in width and up to a mile or two in depth. It probably results from the damming up of the upper-air antitrade winds owing to deflection caused by the earth's rotation. The subtropical jet stream appears to influence the movement of air masses and associated cyclones and anticyclones, which, in turn, greatly affect weather conditions at the surface. It is known to migrate with the seasons and reaches its farthest poleward location in the summer. It is better developed, and thus a greater control of weather conditions, in the winter season.

In addition to the well-developed subtropical jet streams, two other jet streams in each hemisphere have been identified. The polar front jet stream is a west-to-east movement of air, usually poleward of latitude 45°, which is best developed and strongest in the winter season. The polar night jet stream is found at very high altitudes, also flowing from west to east, in the vicinity of the Arctic and Antarctic Circles. There appears to be a correlation between the irregular waves of these jet streams and surface weather changes.

As has been noted, the subtropical jet stream creates ridges of high pressure at about latitude 30° in both hemispheres. These ridges are broken up by land masses into cells which remain well developed only over the ocean areas. From these cells of high pressure, surface air flows outward in a clockwise direction in the Northern Hemisphere and in a counterclockwise direction in the Southern Hemisphere owing to the deflective effect of the earth's rotation. These high-pressure whirls exist over the North and South Atlantic, the North and South Pacific, and the South Indian Ocean (Fig. 3–12). There is no high-pressure whirl in the North Indian Ocean, owing to the existence of the Asian continent at this latitude.

Where the oceanic whirls of opposite hemispheres meet, there is a zone of ascending air and heavy rainfall along the intertropical convergence zone. The whirls have a greater tendency to be in direct contact on the western side of oceans, whereas on the eastern side there is a zone where calm conditions frequently exist. From these relatively quiet areas, local erratic, feeble movements of surface air may flow eastward; these are known as the equatorial westerlies.

FIG. 3-11. Meanders in the jet stream. The jet stream separates cold polar air and warm semitropical air. It meanders widely, and goes through a cycle from minimum to maximum undulation over a period of several weeks. As the meanders develop in the stream, cold air is carried equatorward and warm air poleward (c). Toward the end of the cycle, the meanders become divorced, leaving cells of air displaced considerably from their normal latitude (d). When the jet stream is relatively straight, weather conditions tend to be stable and constant; when the stream oscillates, weather conditions are highly unsettled.

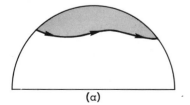 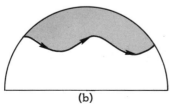 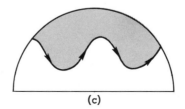 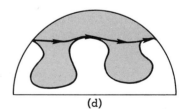

(a)   (b)   (c)   (d)

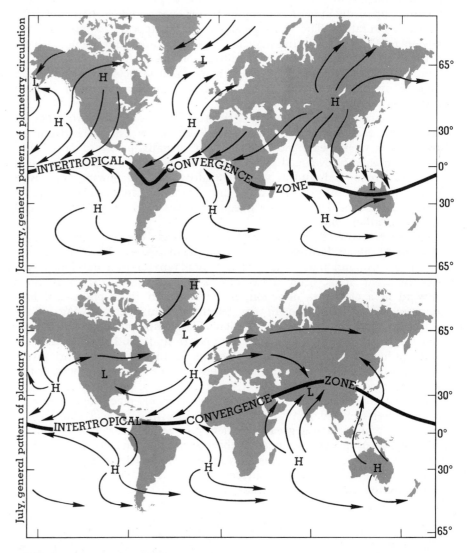

FIG. 3-12. Generalized pattern
of the prevailing surface
winds of the world.

EARTH SCIENCE

During the winter, cold air moves equatorward from the ice caps of the Antarctic, Greenland, the ice-covered Arctic Ocean, and snow-covered continents. The surges of cold air are commonly referred to as polar outbursts and are believed to be associated with irregular waves of the polar front jet stream.

Monsoonal system of winds. The term monsoon is derived from an Arabic word meaning seasonal reversal of winds. In certain areas, winds in the summer blow from approximately the opposite direction from winds in the winter. This seasonal reversal of circulation is best developed in Asia south of the Himalaya Mountains and also exists, though in considerably less developed form, in the southeastern United States, western Africa, northern Australia, eastern Asia, and the Iberian peninsula.

The monsoons of different areas of the world occur for somewhat different reasons; but in all cases they develop only where there is a significant seasonal change in pressure. The outstanding example is India and adjacent southern Asia, where a moist prevailing southwest wind blows from the Indian Ocean over land in the summer. The southwest summer monsoon is associated with heavy rainfall. In the winter season, the winds are dry and northeast in direction, flowing from land to the ocean. This reversal of winds has traditionally been explained as a direct result of seasonal heating and cooling of the continent, with subsequent pressure changes. Increased knowledge of the subtropical jet stream has indicated that the seasonal shifts of the jet stream are a more likely cause of the monsoon of southern Asia.

It is now known that, in the winter season, the subtropical jet stream is forced south of the Himalaya Mountains by the strong continental high-air-pressure area. The jet stream and associated high pressure give rise to equatorward movements of air, which are deflected and become northeast winds. In the summer, the jet stream is well north of the Himalaya Mountains, allowing moist southwest winds to move over the continent.

Along the east coasts of Asia and North America, in the winter season, frequent polar outbursts merge with the general pattern of westerlies winds, resulting in persistent westerly movement of air off land. In the summer, maritime air moves inland from the high-air-pressure whirls over the adjacent oceans.

Local terrestrial winds. In coastal regions, land and sea breezes blowing in a diurnal reversal of winds are often a local characteristic. During the day, a local low pressure develops over the heated land surface and generates a surface breeze from sea to land. At night the land is cooler than the adjacent water and a land-to-sea breeze is the consequence. Sea breezes usually begin during the late forenoon and reach their maximum

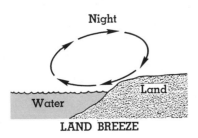

LAND BREEZE

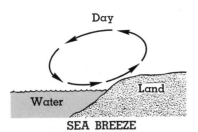

SEA BREEZE

FIG. 3-13. Land and sea breezes are responses to local pressure conditions. Land heats and cools much more than water and thereby becomes a high-air-pressure area at night and a low-air-pressure area during the day. Water remains at about the same temperature and pressure and is a low or high only in relation to air over the land surface.

velocity by the middle or late afternoon. In midlatitudes maximum velocity is 10 to 15 miles per hour, but in tropical areas the sea breeze may be considerably stronger, sometimes reaching storm intensity and usually bringing the benefit of a drop in temperature of 15–20°. Sea breezes may blow inland from 10 to 30 miles, depending upon topography, but are relatively shallow—usually only 800 to 1200 feet in depth. The land breeze is generally weaker, reaching maximum velocities of 8 to 12 miles per hour just before dawn, and it rarely extends more than 5 to 6 miles over the sea. Land and sea breezes are the result of the heating and cooling differential of land and water (Fig. 3–13).

In areas of rugged topography, the development of mountain and valley breezes is not uncommon. The mountain breeze is a nighttime phenomenon resulting from the downslope movement of cool air chilled by terrestrial radiation. It is a form of air drainage and may help develop strong temperature inversions. The valley breeze is an upslope movement of air during the day. A valley floor or slope is likely to become intensely heated because it may receive a more direct ray of the sun. The heated air expands, and there is a rising of air along the valley slopes. In the Northern Hemisphere the southern slopes, because of the stronger insolation, are the sites of well-developed upslope breezes. Valley breezes usually begin in midmorning and last until sunset, at which time the downslope or mountain breeze begins.

In high plateau areas, especially during the winter season, quiet air may become very cold by radiation cooling. The cold, dense air may then flow down the slope of the plateau escarpment, giving rise to a gravity or *katabatic* wind. This downslope movement of cold air under the pull of gravity is common along the margins of the ice caps in Greenland and Antarctica and often reaches gale intensity. In Europe the katabatic winds are known by various names. Along the Adriatic coast, air descending from the adjacent plateau is called the *bora*, and in southern France, the *mistral* is the cold air coming down from the snow-covered interior along the Mediterranean coast. Katabatic winds are best developed in midlatitudes in the winter season and are most common along coasts or lowlands adjacent to plateau areas.

MOISTURE

Although it is but a small percentage of the total, water vapor is the most important constituent of the atmosphere with regard to weather and climate. In some respects it is unique among the gases of the atmosphere. It is the most variable, fluctuating considerably in amount, and it is the only one that is commonly found at normal atmospheric temperatures in all three forms—gas, liquid, and solid. As we have seen, it contributes significantly to the heating and cooling of the atmosphere, and it is the controlling factor in all condensation and precipitation processes.

The amount of water vapor in the air at any given time may be expressed in several different ways:

Absolute humidity is the weight of vapor in a given volume of air (grams per cubic foot or meter). The absolute humidity changes (1) as water vapor is added to or subtracted from the air and (2) as the air expands or contracts. Because it is affected by changes in pressure, absolute humidity is not a good indicator of moisture content.

Specific humidity is the weight of vapor per unit weight of air including water vapor (grams per kilogram). It remains constant regardless of the change in the volume of the air in ascent or descent.

Relative humidity is the amount of water vapor in the air compared with the maximum amount that could be contained at the given temperature. It is always expressed in the form of a fraction, or percentage. For example, at 70° F., a cubic foot of air could hold 8 grains of water vapor if completely saturated. If there are 2 grains per cubic foot present, the air is one-quarter saturated and the relative humidity is 25 per cent. Relative humidity changes as the moisture content changes and also as the air changes temperature, for warm air can contain considerably more water vapor than cold air (Table 3–2). Note that if air at 30° F. is heated to 40° F., its water vapor capacity increases by only 1.0 grains; but if air at 90° F. is heated to 100° F., its capacity increases by 5.0 grains. In the summer, air frequently holds seven to eight times more vapor than during the winter season.

CONDENSATION

As unsaturated air is cooled, its capacity to hold water vapor decreases and eventually reaches a point where the relative humidity is 100 per cent (Table 3–2). Additional cooling beyond that point causes some of the water vapor to change to liquid or solid form, or condense. The *dew point* is the temperature at which condensation begins. It can be ascertained by the following simple experiment. If water is placed in a small metal container and kept well stirred, the container will attain the same temperature as the water. If ice is then added to the water and the water stirred with a thermometer, the container will eventually cool the surrounding air to or below the dew point, and condensation will occur. When small beads of liquid appear on the outer side of the container, the dew point has been reached, and a reading of the thermometer used to stir the water will indicate the dew point temperature.

TABLE 3–2 WATER VAPOR CAPACITY OF 1 CUBIC FOOT OF AIR

TEMPERATURE (°F)	WATER VAPOR (grains)
30°	1.9
40°	2.9
50°	4.1
60°	5.7
70°	8.0
80°	10.9
90°	14.7
100°	19.7

Water vapor can be converted to liquid or solid form only by reducing the temperature of the air to or below dew point. The relative humidity determines how much cooling is necessary before condensation can begin. If the dew point is considerably lower than the actual temperature of the air, a great deal of cooling is necessary before dew point is reached.

Condensation at or near the surface of the earth. Dew, white frost, and fog are forms of condensation that appear at or near the earth's surface. They are common when there are strong static temperature inversions. A long night, clear sky, and relatively calm conditions allow terrestrial energy to escape. The surface of the ground then becomes cool and chills the lower layer of air below the dew point. If the dew point is above 32° F., condensation is most likely to occur in the form of microscopic droplets of water, or dew; if the dew point is below 32° F., condensation will generally take place as small ice crystals, or white frost.

If air is free of solid particles, it may be cooled below the dew point without the occurrence of condensation. The air is then said to be supersaturated. If, on the other hand, there is an abundance of hygroscopic nuclei or smoke or salt particles, condensation may begin at relative humidities well below 100 per cent. The presence of great numbers of microscopic hygroscopic particles over coasts and cities explains the frequency and density of fogs in these localities. It is also not uncommon for water vapor to change into liquid droplets in air above the ground even when temperatures are below 32° F. Supercooled droplets may exist in a cloud or fog where air temperatures are as low as −40° F. There is no ready explanation for the existence of these supercooled droplets. It is most likely their colloidal stability that accounts for their presence, for if broken up by a moving vehicle or aircraft, they rapidly freeze into ice.

Fog consists of microscopic droplets of water gathered in sufficient numbers to impede visibility. It may result either from the cooling of air near the surface by a cold ground or from the addition of water vapor to the atmosphere and the consequent increase in the dew point to equal the actual air temperature.

Radiation fog, or ground fog, is very common. It results from cooling of the night air over a cold surface by radiation and conduction. This occurs when strong temperature inversions are present and the air is relatively humid. Often cold air that has accumulated in valleys by air drainage is chilled below the dew point, causing fog to develop in the lowlands while the higher lands remain clear. Radiation fogs develop from the surface upward as the radiation cooling continues through the night. They dissipate in the morning when the sun rises.

Advection fog, another common type, is produced when warm, moist air blows over a cold surface and the colder surface lowers the temperature of the air to the point of condensation. Advection fogs occur when warm, moist air over the ocean is blown over a cold land surface—usually in the winter season—or when the air over warm water is transported over a cold ocean current.

Frontal fogs form in the winter season in midlatitudes along the boundaries of two contrasting air masses. Here warm, moist air is forced over colder, denser air, and rain may result; the raindrops falling through the cold-air wedge may then evaporate, adding moisture to the lower air and thereby raising its dew point to the actual temperature of the air.

Dense fogs are a serious menace to air, land, and water transportation. Thus far, fog dispersal has been possible only within a very local area and at high cost. Three methods have been used on an experimental basis. One is to heat the air to a temperature above its dew point; another is to dry the air by means of chemical materials with water-absorbing qualities; and the third makes use of electrically charged materials that are dropped through the fog. Despite these attempts, it is unlikely that any means will be found to clear fog over large areas at low cost.

Condensation aloft. Clouds are made up of small droplets of water and/or ice crystals. They are the result of rising air that is cooled by expansion under reduced pressure until dew point is reached and condensation occurs. Ascending air cools at a uniform rate regardless of the temperature of the air as long as condensation does not take place. This rate is approximately 5½° per 1000 feet, and it is known as the *dry adiabatic rate of cooling*. It differs from the lapse rate in that it applies to a body of air that is moving vertically. The lapse rate is the decrease of temperature through different layers of air that are not moving vertically. The lapse rate averages 3.5° per 1000 feet but may vary from zero, or a negative rate in a temperature inversion, to over 19° per 1000 feet in very turbulent air. Air that is descending heats at a uniform rate of 5½° per 1000 feet, and this is known as the heating-by-compression rate or *adiabatic heating rate*. It is important to distinguish clearly between the lapse rate, the dry adiabatic rate, and the heating-by-compression rate.

We have noted that the rate of cooling of ascending air does not change as long as condensation does not take place. But rising air may cool to the point where condensation does occur. Then latent heat is released and the rate of any subsequent cooling is retarded. The slower rate of cooling of saturated air is called the *wet* or *retarded adiabatic rate*, and it varies depending upon the amount of water vapor in the air. In relatively cold, dry air, little condensation takes place, and the wet adiabatic rate may be as high as 3½–4½° per 1000 feet. In warm and moist air, a great amount of latent heat is released, greatly reducing the cooling process, and the wet adiabatic rate may be as low as 2–2½° per 1000 feet.

The relationship of the existing lapse rate to the adiabatic rates determines whether the air is stable or unstable. Stable air resists any vertical movement and if forced aloft will tend to return to its original position; it is associated with clear and dry weather conditions. If the lapse rate is lower than 2½° per 1000 feet, a condition of *absolute stability* exists. Such air resists vertical displacement under all circumstances. If, on the other hand, the lapse rate is greater than the dry adiabatic rate of 5½° per 1000 feet, *absolute instability* prevails. Air forced to rise will continue to ascend because its cooling rate does not equal the temperature of the air surrounding it. Unstable air is subject to vertical motion and is often associated with clouds and precipitation. Lapse rates between 2½° and 5½° produce stable conditions if the air is dry and little latent heat is released, but if the air is moist and a great deal of latent heat is released, *conditional instability* prevails; the "condition" referred to is the condensation of large amounts of water vapor with the consequent release of heat and resulting instability. It can be seen, therefore, that (1) low lapse rates produce stable air with little if any vertical motion, and (2) the higher the lapse rate, the more turbulent and unstable the air. With a lapse rate of 19° or more, very violent conditions exist. Under such conditions of *mechanical instability*, air will automatically overturn without any "triggering," or forced displacement. Such lapse rates occur in violent storms such as tornadoes.

Clouds reflect the processes that are taking place in the atmosphere and are good indicators of wind and moisture conditions and of the stability of the air. For this reason, they are useful in weather forecasting. The great variety of clouds can be combined into 10 principal types now recognized internationally (Fig. 3–14). The high clouds, found between 20,000 and 35,000 feet, are thin and wispy and are composed of ice crystals. They are called *cirrus* (Ci), *cirrostratus* (Cs), and *cirrocumulus* (Cc), in order of descending elevation (Figs. 3–15 to 3–17). Cirrus clouds are featherlike in appearance and, when arranged in a pattern, usually indicate approaching bad weather. Cirrostratus clouds are thin and usually cover the entire sky, giving it a whitish appearance. Often the ice particles making up these clouds refract light, producing a halo around the moon or sun. Cirrocumulus clouds are arranged in ripples and are often referred to as a mackerel or buttermilk sky.

The middle-cloud family consists of *altostratus* (As) and *altocumulus* (Ac) clouds, found at elevations of 6500 to 20,000 feet (Figs. 3–18 and 3–19). Altostratus clouds have the appearance of a high, gray sheet, usually fibrous in structure. The sun or moon can be seen through them only faintly.

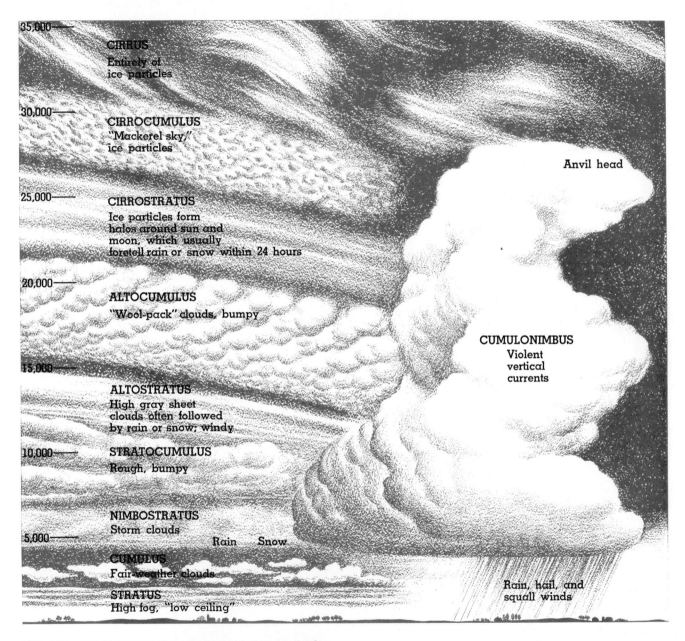

Labels within figure:

35,000 —

CIRRUS
Entirely of
ice particles

30,000 —

CIRROCUMULUS
"Mackerel sky,"
ice particles

25,000 —

CIRROSTRATUS
Ice particles form
halos around sun and
moon, which usually
foretell rain or snow within 24 hours

20,000 —

ALTOCUMULUS
"Wool-pack" clouds, bumpy

15,000 —

ALTOSTRATUS
High gray sheet
clouds often followed
by rain or snow; windy

10,000 —

STRATOCUMULUS
Rough, bumpy

NIMBOSTRATUS
Storm clouds

5,000 —

Rain   Snow

CUMULUS
Fair-weather clouds

STRATUS
High fog, "low ceiling"

Anvil head

CUMULONIMBUS
Violent
vertical
currents

Rain, hail, and
squall winds

FIG. 3-14. Generalized vertical arrangement of major cloud
types. (Redrawn from A. J. Nystrom & Company with permission.)

FIG. 3-15. High-level cirrus (Ci) clouds. Cirrus clouds with a definite pattern may indicate the invasion of a moving low air pressure and subsequent changing weather. (Environmental Science Services Administration, U.S. Department of Commerce.)

FIG. 3-16. High-level cirrostratus (Cs) clouds. These clouds are most commonly found at elevations of 20,000 to 25,000 feet, covering the entire sky. The ice particles making up cirrostratus clouds frequently refract light, producing a halo around the moon or sun. (Don Whelpley and Environmental Science Services Administration, U.S. Department of Commerce.)

The low clouds, under 6500 feet, consist of the *stratocumulus* (Sc), *stratus* (St), and *nimbostratus* (Ns) types (Figs. 3–20 and 3–21). Stratocumulus clouds appear as a gray overcast with large, semicircular, thicker masses interspersed. Stratus clouds are low, gray clouds producing general overcast conditions. Nimbostratus clouds are low, widespread clouds associated with continuous drizzlelike precipitation.

Finally, there are two types of clouds that are very thick, often reaching a vertical development of over 20,000 feet. These are the *cumulus* (Cu) and the *cumulonimbus* (Cb) clouds (Figs. 3–22 and 3–23). Cumulus clouds appear as large cotton balls and are very common in the summer season. Cumulonimbus clouds are very thick thunderheads always associated with thunderstorms and often with violent conditions.

The ten cloud types are actually combinations of four basic clouds:

Cumulus—thick, flat-based, and dome-shaped

Stratus—low; create overcast skies; resemble high-level fog

Cirrus—high, whispy, and featherlike

Nimbus—rain-bearing clouds

As we have seen, clouds are produced by the condensation of rising and cooling air. They are the visible portions of ascending currents. The forced ascent of large volumes of air with attendant cooling and the

FIG. 3-17. Cirrocumulus (Cc) clouds. These clouds are characterized by closely packed, small globular masses often arranged in lines or ripples. (Don Whelpley and Environmental Science Services Administration, U.S. Department of Commerce.)

FIG. 3-18. Altostratus (As) clouds. These clouds usually cover the entire sky at elevations of 10,000 to 15,000 feet, giving the appearance of a thin gray sheet. The sun or moon can be seen through them only faintly. (Don Whelpley and Environmental Science Services Administration, U.S. Department of Commerce.)

FIG. 3-19. Altocumulus (Ac) clouds. Notice banding effect, which is not uncommon at levels of 12,000 to 20,000 feet. (Environmental Science Services Administration, U.S. Department of Commerce.)

▲

FIG. 3-20. Stratocumulus (Sc) clouds. These clouds occur in layers or patches of globular masses. Often the edges of the rolls join together to make a continuous cloud cover with a wavy appearance. (U.S. Department of Agriculture.)

FIG. 3-21. Stratus (St) and ▶ nimbostratus (Ns) clouds. Low-level clouds associated with widespread overcast and drizzle rain or snow. (U.S. Department of Agriculture.)

AIR PRESSURE, WINDS, AND MOISTURE

FIG. 3-22. Fair-weather cumulus (Cu) clouds. Note the flat base and vertical development. (Don Whelpley and Environmental Science Services Administration, U.S. Department of Commerce.)

EARTH SCIENCE

FIG. 3-23. Cumulonimbus (Cb) clouds. Showers can be seen falling from the base of the cloud. The characteristic anvil shape has not yet developed. (U.S. Department of Agriculture.)

possibility of condensation and precipitation is accomplished by three processes, which may occur singly or in combination:

1. Convection—the rising of air heated by a warm earth surface.
2. Orographic ascent—the rising of air over land barriers perpendicular to the path of prevailing winds.
3. Cyclonic wedging—the rising of warm, moist air over cool, dense air.

Figure 3–24 illustrates these three processes.

FIG. 3-24. Major types of forced air ascent.

Convectional: Air warmed by the earth's surface expands and rises.

Warm air

Surface

Orographic: Prevailing winds are forced to rise over a topographical barrier.

Prevailing wind

Surface

Cyclonic: Warm, moist air is wedged over cold, dry, and denser air.

Warm air

Cold air

Ascending air may condense and form clouds without yielding precipitation. For precipitation to take place, the droplets of water or ice particles in a cloud must be of sufficient size so that they fall, rather than being held up by rising air currents or evaporating on descent. Which of the various forms of precipitation occurs depends upon the temperature at which condensation takes place, the turbulence of the air, and the type and number of hygroscopic nuclei in the cloud.

Rain. Precipitation that falls to the surface as a liquid is called rain. This is the most common of all types of precipitation over the earth as a whole. There are two processes by which small cloud droplets may be combined into drops large enough to fall. The first and most important requires the coexistence of water droplets and ice particles within a cloud. The ice particles serve as nuclei around which supercooled droplets of water accumulate and finally reach a size at which they fall to earth. If an appreciable amount of rain is to fall, the ascending air must rise above the level of freezing, where there is direct conversion of water vapor to solid form (sublimation), and where supercooled droplets of water also exist. The seeding of clouds with silver iodide is based on the principle of providing hygroscopic nuclei around which water droplets may gather. Silver iodide particles are known to be good hygroscopic nuclei and are therefore used as an inexpensive substitute for ice crystals.

The second process by which raindrops may form is the coalescence of water droplets as they move about violently in turbulent air. Through collisions, droplets merge into single drops large enough to fall. This is possible only in turbulent air such as exists within a cumulonimbus cloud. In relatively quiet air, raindrops can grow as large as 0.2 inch in diameter, but they break into small droplets at velocities of about 18 miles per hour. Hence, raindrops cannot descend as precipitation through vertical currents of this velocity or more.

Snow. Where temperatures are well below freezing, water vapor may change directly to ice (*sublimation*), and precipitation may occur in the solid stage. Snowflakes are ice crystals, always hexagonal in form but with a great variety of intricate patterns. Large snowflakes are formed by the coalescence of many small individual crystals. Cold air contains little moisture, and for this reason precipitation is unlikely to occur, or will be very light, on bitterly cold days. Conversely, heavy snowfall occurs when temperatures at the level where crystals may form are not much below freezing. As previously noted, snow is a poor conductor of heat, and a snow-covered surface will keep the *ground* temperature higher than it otherwise would be. On the other hand, *air* temperature will be considerably lower because of the increased reflection of solar energy during the day, and because snow acts as a blanket in retarding heat transfer from ground to air at night.

Figure 3–25 shows an instrument widely used to measure and record the amount of precipitation in the form of rain or snow.

Sleet. Raindrops may freeze or partially freeze as they fall through the air. This occurs when a strong temperature inversion exists above the surface and warm, moist air is wedged above a layer of subfreezing air. Sleet is precipitation that begins to fall as rain but becomes frozen or partially frozen before reaching the earth.

Hail. Hard, circular pellets composed of concentric layers of clear and opaque ice are known as hail. Individual hailstones have been known to reach a size of several inches in diameter. One explanation for their formation is that they begin as droplets of water or ice particles and are lifted by vertical currents into the upper cloud, where it is well below freezing. Here, they acquire a layer of snow or opaque ice. They then fall into the lower and warmer parts of the cloud, accumulating a coating of clear ice or water. Eventually they are carried upward again into the icing area and gather another layer of milky ice, descend, and collect a coating of water or clear ice, and so on. This repeated bouncing accounts for the layered structure of hailstones. Hail is produced only from cumulonimbus clouds, which are characterized by vigorous ascending and descending currents of air. An alternative theory of the formation of hailstones suggests that they are formed by falling through air in which there are alternate layers of snow and supercooled droplets of water and thus acquiring alternate coatings. Large hailstones may be very destructive to growing crops, buildings, and even livestock. Hailstorms are most frequent in the central United States, and during the spring and early summer.

Glaze and rime. Glaze and rime are not forms of precipitation but rather unusual accumulations of ice on features and objects on or near the surface. The term glaze refers to ice that forms when supercooled raindrops freeze as they break up upon striking solid objects. During an ice storm, glaze—or clear ice—may accumulate in deposits more than 2 inches in diameter. The weight of this ice may be sufficient to break limbs of trees and snap electrical wires (Fig. 3–26). The slippery surface conditions produced are a serious hazard to motorists, airplanes, and pedestrians.

Rime is ice that accumulates in layers on the windward side of buildings, trees, poles, and other objects. Supercooled droplets existing in a fog or mist are blown against a solid surface, where they break and immediately freeze. Deposits of rime, or layered ice, may accumulate to

FIG. 3-25. Rain- and snow-weighing gauge. The accumulated weight of water or snow caught in the cylinder moves a platform supported by a spring balance. This movement is communicated to a pen which writes a continuous record on a clock-driven drum. The record reads in inches of precipitation. (Environmental Science Services Administration, U.S. Department of Commerce.)

FIG. 3-26. The effects of glaze and ice. The weight of ice has been known to snap electrical wires and break limbs of trees. Severe ice storms have caused transportation to come to a halt for periods of several days. (Environmental Science Services Administration, U.S. Department of Commerce.)

several feet on the windward side of objects, under ideal conditions. Rime is most frequently found in mountainous regions where fog and mist are common.

SIGNIFICANT PRECIPITATION DATA

Since water is one of the necessities of life, the rainfall characteristics of a locality comprise a very important aspect of its physical environment. The total annual average, and its seasonal distribution, are among the most significant items of precipitation information.

Annual rainfall. The annual amount of rainfall is expressed in inches of water per year. Usually, 1 foot of snow equals about 1 inch of rain, but this ratio can vary greatly, depending upon the density of the snow. Differences in the annual amount of rainfall from one place to another depend upon proximity to the oceans, which are the ultimate source of all precipitation. Several areas in the world average considerably less than 1 inch, others receive over 400 inches in the course of an average normal year (Fig. 3–27). The map of average annual rainfall indicates the following general patterns.

1. Rainfall is greatest in the equatorial regions and decreases toward the poles.
2. Rainfall is heaviest in coastal regions and decreases toward the interior of continents.
3. Rainfall is very heavy on the windward side of highlands; very dry conditions prevail on the leeward side.
4. Coastal areas adjacent to cold ocean currents are drier than coastal areas near warm currents.
5. In general, humid conditions prevail in the equatorial low, westerly winds, and subpolar lows.
6. In general, dry conditions prevail in the polar highs, trade winds, polar easterlies, and subtropical highs.

Seasonal distributions. The season or seasons in which rainfall occurs in any specific locality is of major importance to its welfare. This is especially true in midlatitudes, where a cold season precludes plant growth during part of the year, so that only the rainfall that occurs during the frost-free period is useful for agricultural purposes. The seasonal occur-

FIG. 3-27. Generalized map of the distribution of average annual precipitation.

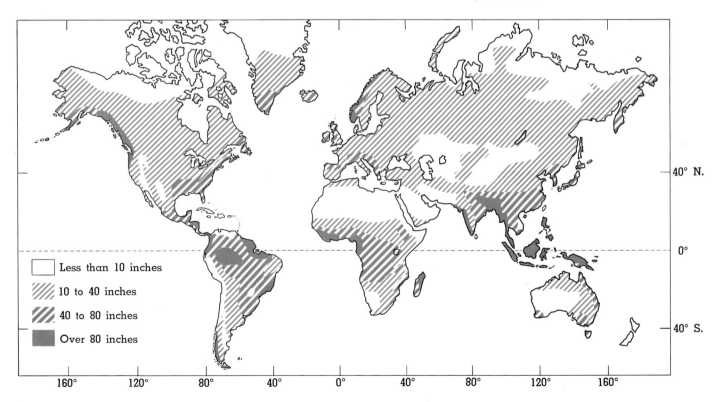

rence of precipitation varies greatly over the earth. In some areas the annual rainfall is well distributed over both seasons, whereas in other areas it is concentrated in one period of the year. Seasonal rainfall may be produced by either of two conditions:

1. The annual migration of the winds and pressure belts results in the control of moisture conditions by a dry wind or pressure belt during one season and by a humid belt during the other season.
2. The seasonal reversal of winds produces a dry, land-derived monsoon during one season and a moist, sea-derived monsoon in the other.

Other significant precipitation data include the annual and seasonal deviations from the average. Rainfall usually varies from one year to another in a specific locality. Generally, the deviation from the mean, or average, is greater in dry areas, where the average total amount is low, and is relatively slight in humid areas having heavy rainfall. In semiarid areas, where there is a degree of risk in any agricultural venture, the deviation from the mean is of utmost importance. Also of importance is the intensity of rainfall—the number of days of rain in relation to the total amount. The intensity of rainfall differs greatly over the earth. For example, during a normal year some areas in southern England receive about 25 inches of rain spread over 160–180 days. The rainfall is cyclonic in origin for the most part and is characterized by low, overcast nimbostratus clouds and a slow rate of drizzle. On the other hand, areas in India adjacent to the Himalayan Mountains receive about 430 inches of rain a year in 150–160 days. Most of the precipitation occurs in the summer season, in the form of fast, sharp showers associated with the moist summer monsoon that is forced to rise over the mountains. The winter season is considerably drier, as the land-derived winter monsoon blows over this region.

SUMMARY OF PROCESSES ACTIVE IN THE ATMOSPHERE

- High angles of the sun's ray result in high temperatures; low angles are associated with low temperatures. Hence, tropical areas are warm throughout the year and polar areas are cold at all seasons.
- The vertical ray of the sun migrates latitudinally over the course of the year. As a result, in midlatitudes high angles and a warm season occur for part of the year and low angles and a cold season for the other part.

- The length of day varies everywhere except at the equator. In the midlatitudes long days coincide with high angles of the sun and warm seasons. Short days coincide with low angles and cold seasons.

- Land heats and cools more than do water bodies when exposed to an equal amount of solar radiation. Consequently, extremes of temperatures are found in the interiors of large land masses.

- The earth's surface, rather than the sun, is the primary direct source of heat for the atmosphere. Hence, temperatures normally decrease aloft, averaging a drop of 3½° per 1000 feet (lapse rate).

- Under certain conditions, cold air may rest at the surface with warm air above. Temperatures then increase with increasing elevation in the first few hundred feet of the atmosphere (temperature inversions).

- Air pressure decreases rapidly aloft, dropping an average of about 1 inch per 1000 feet in the first few thousand feet above sea level.

- At the surface, high temperatures are normally associated with low-air-pressure centers and low temperatures with high air pressures. Extremes of surface air pressures are therefore found in the interior of large land masses.

- Winds blow from high-pressure centers toward low-pressure areas. The greater the difference between the high and low pressure zones (pressure gradient), the greater the velocity of the wind.

- Low-pressure areas are zones of converging winds and ascending currents; high-pressure areas are zones of diverging winds and descending currents.

- Winds, like all other moving bodies, have an apparent deflection to the right in the Northern Hemisphere and to the left in the Southern Hemisphere as a result of the earth's rotation.

- Warm air can contain considerably more water vapor than cold air.

- Water vapor condenses to liquid or solid form when air is cooled sufficiently.

- Large volumes of air can be cooled only by being forced to rise. Nonsaturated ascending air cools at a uniform rate of 5½° per 1000 feet (dry adiabatic rate). Descending air heats at a uniform rate of 5½° per 1000 feet (adiabatic heating rate).

- Dew point is the temperature at which water vapor condenses. The greater the difference between dew point and the actual temperature of the air, the greater the amount of cooling necessary to reach the condensation stage. The dew point is often considerably lower than the actual temperature of unsaturated air.

- Relative humidity decreases as the temperature rises and increases as the temperature drops.

TABLE 3-3 DEW POINT TEMPERATURE (°F)[a]

AIR TEMPERATURE (°F)	DEPRESSION OF THE WET-BULB THERMOMETER													
	1	2	3	4	6	8	10	12	14	16	18	20	25	30
0	−7	−20												
5	−1	−9	−24											
10	5	−2	−10	−27										
15	11	6	0	−9										
20	16	12	8	2	−21									
25	22	19	15	10	−3	−15								
30	27	25	21	18	8	−7								
35	33	30	28	25	17	7	−11							
40	38	35	33	30	25	18	7	−14						
45	43	41	38	36	31	25	18	7	−14					
50	48	46	44	42	37	32	26	18	8	−13				
55	53	51	50	48	43	38	33	27	20	9	−12			
60	58	57	55	53	49	45	40	35	29	21	11	−8		
65	63	62	60	59	55	51	47	42	37	31	24	14		
70	69	67	65	64	61	57	53	49	44	39	33	26	−11	
75	74	72	71	69	66	63	59	55	51	47	42	36	15	
80	79	77	76	74	72	68	65	62	58	54	50	44	28	−7
85	84	82	81	80	77	74	71	68	64	61	57	52	39	19
90	89	87	86	85	82	79	76	73	70	67	63	59	48	32
95	94	93	91	90	87	85	82	79	76	73	70	66	56	43
100	99	98	96	95	93	90	87	85	82	79	76	72	63	52

[a] Barometric pressure, 30.00 inches.

- Rising air that is saturated and condenses continues to cool as it ascends but at a rate lower than 5½° per 1000 feet (wet adiabatic rate), owing to the addition of latent heat of condensation. The warmer and wetter the air, the more latent heat released and the slower the rate of cooling.

- Clouds and precipitation are normally associated with low-pressure areas; dry and fair weather conditions are associated with high-pressure areas.

- The higher the lapse rate, the less stable the air is likely to be; conversely, the lower the lapse rate, the more stable the air.

- The dew point decreases about 1° per 1000 feet of lift of unsaturated air.

- Warm, humid air is associated with potentially unstable conditions; cold, dry air tends to be stable.

TABLE 3-4 RELATIVE HUMIDITY (%)[a]

AIR TEMPERATURE (°F)	DEPRESSION OF THE WET-BULB THERMOMETER													
	1	2	3	4	6	8	10	12	14	16	18	20	25	30
0	67	33	1											
5	73	46	20											
10	78	56	34	13										
15	82	64	46	29										
20	85	70	55	40	12									
25	87	74	62	49	25	1								
30	89	78	67	56	36	16								
35	91	81	72	63	45	27	10							
40	92	83	75	68	52	37	22	7						
45	93	86	78	71	57	44	31	18	6					
50	93	87	80	74	61	49	38	27	16	5				
55	94	88	82	76	65	54	43	33	23	14	5			
60	94	89	83	78	68	58	48	39	30	21	13	5		
65	95	90	85	80	70	61	52	44	35	27	20	12		
70	95	90	86	81	72	64	55	48	40	33	25	19	3	
75	96	91	86	82	74	66	58	51	44	37	30	24	9	
80	96	91	87	83	75	68	61	54	47	41	35	29	15	3
85	96	92	88	84	76	70	63	56	50	44	38	32	20	8
90	96	92	89	85	78	71	65	58	52	47	41	36	24	13
95	96	93	89	86	79	72	66	60	54	49	44	38	27	17
100	96	93	89	86	80	73	68	62	56	51	46	41	30	21

[a] Barometric pressure, 30.00 inches.

- The windward side of highlands receives heavy precipitation; leeward sides are dry.
- The oceans are the primary source of moisture. The amount and character of precipitation in an area depend on its distance from the ocean and the existence of intervening topographical barriers.
- Water does not necessarily freeze at 32° F. Supercooled droplets may exist in a fog or cloud at temperatures well below zero.
- The difference between dry-bulb (actual air temperature) and wet-bulb temperatures determines the dew point and the relative humidity of the air. In any given set of observations, the greater the difference, the drier the air and the lower the dew point and the relative humidity. (Tables 3-3 and 3-4 give humidity values and psychrometric observations that may be used to determine dew points and relative humidities. These tables are condensed from the *Smithsonian Meteorological Tables* published by the Smithsonian Institution.)

SUGGESTED REFERENCES _____

Battan, L. J.: *Cloud Physics and Cloud Seeding*, Doubleday & Company, Inc., Garden City, N.Y., 1962.

Blair, T. A., and R. C. Fite: *Weather Elements*, Prentice-Hall, Inc., Englewood Cliffs, N.J., 1965.

Flora, S. D.: *Hailstorms of the United States*, University of Oklahoma Press, Norman, Okla., 1956.

Hare, F. K.: *The Restless Atmosphere*, Hutchinson and Company, Ltd., London, 1953.

International Cloud Atlas, World Meteorological Organization, Geneva, Switzerland, 1956.

Mason, J. B.: *Clouds, Rain and Rainmaking*, Cambridge University Press, New York, 1962.

Perrie, D. W.: *Cloud Physics*, University of Toronto Press, Toronto, 1951.

Petterssen, S.: *Introduction to Meteorology*, McGraw-Hill Book Company, Inc., New York, 1959.

Riehl, H.: *Introduction to the Atmosphere*, McGraw-Hill Book Company, Inc., New York, 1965.

CHAPTER 4 **AIR MASSES, CYCLONES, AND VIOLENT STORMS**

4

Much of the day-to-day weather in the midlatitudes results from the interplay of large masses of air that have derived their temperature and moisture characteristics by remaining over extensive land or water surfaces. Air masses are moved horizontally from their area of origin by the passing of low- and high-air-pressure cells. These moving lows and highs, or *cyclones* and *anticyclones*, are therefore of outstanding importance in controlling both local and regional weather.

Hurricanes, tornadoes, and thunderstorms have great impact locally, often causing considerable damage, but are relatively unimportant in a climate or on a regional scale.

AIR MASSES

An air mass is an extensive body of air that has approximately uniform temperature and moisture content along a given elevation. Air masses cover several thousand square miles in horizontal extent and are of considerable depth. Their characteristics develop as they remain over a homogeneous surface for a time.

As a result of radiation from the ground and air, and convection and turbulence, an air mass assumes the temperature and moisture conditions of the surface below. Thus, extensive surfaces that are relatively homogeneous impart their characteristics to bodies of air and are known as *source regions*. A simple way of classifying air masses is by reference to their source region—tropical or polar and continental or marine. As one would expect, polar air masses are cold, tropical air masses warm; continental masses are dry, and marine, or maritime, air masses humid.

Air masses do not necessarily remain over their source region but are often moved by passing cyclones and anticyclones. When this occurs air masses of different characteristics are brought together, but they do not mix readily. The boundary between them is called a *front*, or *surface of discontinuity*. Fronts are zones of transition, from 3 to 50 miles in width, within which weather conditions change in an irregular way from those of one air mass to those of another. Weather changes in the passing of a front are considerably more rapid and intense than those within air masses themselves. When a mass moves from its source region to an area with different surface conditions, it gradually loses its original characteristics and takes on those of the new region. The change in its moisture and temperature conditions in turn affects the stability of the air mass. When forced to rise and descend over high mountains, air masses lose their original characteristics very rapidly and are often completely neutralized.

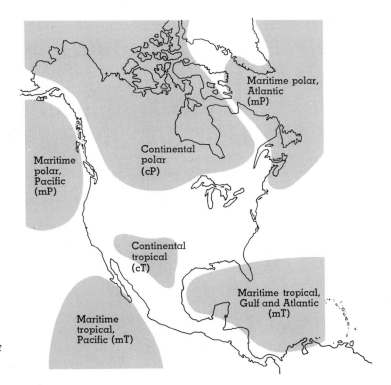

FIG. 4-1. Source regions of the North American air masses.

AIR MASSES OF NORTH AMERICA

The air masses of the North American continent are grouped into six major categories (Fig. 4–1):

Continental polar (cP)
Continental tropical (cT)
Maritime tropical (mT), Pacific Ocean
Maritime tropical (mT), Gulf of Mexico and Atlantic Ocean
Maritime polar (mP), Pacific Ocean
Maritime polar (mP), Atlantic Ocean

The interplay of these air masses greatly influences the day-to-day weather in North America, but no one locality is affected by all of them. As we have said, air masses are greatly modified or neutralized entirely in the course of migrating. Consequently, only two or three types of masses can invade any given area with their original characteristics intact.

Continental polar air mass. Source region: Central Canada, Alaska, northeastern Siberia, the Arctic Ocean.

In this area of ice-covered surfaces, the air in winter becomes very cold and dry. Furthermore, because of the long nights and excessive radiational cooling the lower layers of air become colder than the air above. This causes a steep temperature inversion and a high degree of stability. In summer the ice and snow disappear and the land warms the air to some degree. The air nevertheless remains cool in relation to the air to the south. The continental polar air mass is therefore cold, dry, and stable, especially in the winter. Bitter cold weather in the United States generally, and east of the Rocky Mountains especially, is due to the invasion of cP air.

Continental tropical air mass. Source region: Southwestern United States and northern Mexico.

Because of the narrowness of the continent at these latitudes, an extensive source region is lacking. As a consequence, cT air is of little significance in influencing weather conditions outside its immediate region of origin. This air mass is present only in the summer season and is hot, dry, and unstable. As a result of intense surface heating, convection and turbulence may extend to heights of 2 miles or more.

The only extensive source region of cT air in the Northern Hemisphere is the Sahara area of North Africa. The air mass formed there greatly affects weather conditions in southern Europe at times.

Maritime tropical air mass, Pacific. Source region: Subtropical Pacific Ocean, off the coast of southwestern United States and adjacent Mexico.

The relatively cool water for the latitude makes the Pacific mT air mass appreciably cooler than its counterpart over the Gulf of Mexico and the Atlantic Ocean. Generally, this air mass affects only the southwestern part of the United States and northwestern Mexico. It does not bring as much precipitation as its Gulf counterpart unless it is forced to rise over mountains or a colder wedge of air. When it moves northward into colder areas, the lower layers may be chilled, with fog resulting, especially at night. This air mass is ineffective during summer months, largely because of the strong high air pressures found at latitudes 35° and 40°.

Maritime tropical air mass, Gulf and Atlantic. Source region: Gulf of Mexico and western Atlantic Ocean.

The surface waters here have constantly high temperatures, varying from 70 to 80° F. depending upon the locality, and thus give rise to an exceptionally warm and humid air mass. Temperatures and specific humidities are greater within mT-Gulf air than in any other air mass affecting the United States. This air mass is warm, humid, and unstable and is active in both summer and winter, bringing most of the moisture that exists in the air over the United States and Canada, east of the Rocky Mountains. In summer it is responsible for much of the hot, humid weather as well as most of the thunderstorms and rainfall. In winter it often brings fog as the moist lower layers of air are cooled by the colder land surface; heavy snowfalls; and, occasionally, high temperatures resulting in winter thaws. This air mass, along with the continental polar air mass, is responsible for most of the weather in eastern North America.

Maritime polar air mass, Pacific. Source region: North Pacific Ocean in the vicinity of the Aleutian low-air-pressure area.

The waters of the North Pacific are relatively warm for the latitude; this results in the formation of a warm and humid air mass in the winter, for the land is then considerably colder than the water. When the air is forced over the western mountains, considerable rain or snow may fall along the windward or western slopes. Some localities in this area receive more total precipitation for the year than any other part of the United States. In crossing the western and Rocky Mountains, this air mass is considerably modified, losing its original characteristics. It does not, therefore, markedly influence weather conditions in central and eastern United States and Canada.

In the summer the waters of the North Pacific are cooler than the adjacent land surfaces. Consequently, the mP-Pacific air, as it moves over land, is relatively stable. This air mass affects only the western part of North America. It is warm, moist, and unstable in the winter season and relatively cool and stable in the summer. The mP-Pacific air moves southward along the Pacific Coast in the summer into California, usually preventing mT air from reaching the coast. This, together with the effect of the cool ocean currents south of latitude 40°, gives coastal southern California cool summers and temperatures approximately the same as the Pacific Northwest.

Maritime polar air mass, Atlantic. Source region: North Atlantic Ocean, off the coasts of Newfoundland, Labrador, and Greenland.

This source region is poleward of the Gulf Stream and North Atlantic Drift, and the surface waters are cold. The air mass does not affect the United States appreciably, as it is on the leeward coast and the prevailing westerlies normally push the air eastward toward Europe. Only at times are the cyclones and anticyclones positioned in such a way as to cause mP-Atlantic air to move over the continent. It is unusual for the influence of this air mass to be felt as far south as Virginia, or west of the Appalachian Mountains. In the winter, mP-Atlantic air is moist and cool at its base and relatively dry and stable aloft. Damp, windy, and cool weather, often accompanied by misting rain or snow, is associated with the invasion of mP-Atlantic air. Along the New England coast, this disagreeable weather is usually brought in by northeastern winds and is referred to locally as a *northeaster*. The air mass moves westward over the continent more frequently in summer than in winter and is then associated with cool and fair weather. The mP-Atlantic air was originally cP air blown off the continent and modified. It is cool, moist, and moderately unstable in the winter; and cool, dry, and stable in the summer.

In the upper troposphere, a large volume of warm and dry air, known as *Superior* (S) air, is found occasionally, ranging over much of the United States. Its source is probably in upper air in the vicinity of the subtropical high belt. It normally remains aloft but occasionally appears at the surface, causing very dry, hot weather. Prolonged droughts in the southwestern United States and the Great Plains usually reflect the presence of S air at the surface. All air that is warm and very dry (relative humidity of less than 40 per cent) is ordinarily called Superior air. It is assumed that such warm and very dry air acquires its characteristics by descending from aloft and being compressed and heated in the process.

Table 4–1 gives in condensed form the most important information about the North American air masses.

AIR MASSES OF OTHER PARTS OF THE WORLD

Extensive air masses develop over other continents and adjacent oceans and affect weather conditions greatly. For example, in Europe a cP air mass forms over the Eurasian continent; an mP air mass, over the warm waters of the North Atlantic Ocean; and an mT air mass, over the subtropical eastern vicinity of the Azores high. The hot, dry cT air of the North Africa source region influences the weather in southern Europe appreciably. In Asia, a very well-developed continental polar air mass originates over Siberia and Mongolia; mP air, over the Okhotsk Sea; and an mT air mass, over South China Seas. Similar air masses develop in the Southern Hemisphere and control the weather and climate there.

MIDLATITUDE CYCLONES

Cyclones are moving low-air-pressure cells found chiefly in the westerly wind belts at latitudes 20° to 70° in both hemispheres. They move in a general west-to-east direction at average velocities of 10–20 miles per hour in the summer season and 25–40 miles per hour in the winter. On occasion they may slow down or stagnate completely for a period of a few days; on the other hand, a well-developed storm may move for short distances at velocities as great as 60 miles per hour under very unusual conditions. Cyclones move more slowly in the summer and are less well developed then. Consequently,

T A B L E 4 – 1   CHARACTERISTICS OF THE NORTH AMERICAN AIR MASSES

AIR MASS	SOURCE REGION	CHARACTERISTICS	MAJOR REGIONS OF INFLUENCE
Continental polar (cP)	Alaska, Canada, and the Arctic	Cold, dry, and stable	Eastern North America
Continental tropical (cT)	Southwestern United States and northern Mexico	Hot, dry, and unstable	Southwestern United States, and northern Mexico (summer only)
Maritime tropical, Pacific (mT)	Subtropical eastern Pacific Ocean	Warm, moist, and neutral	Southwestern coast of United States and northwestern Mexico (winter season primarily)
Maritime tropical, Gulf and Atlantic (mT)	Gulf of Mexico, Caribbean Sea, Sargasso Sea, and subtropical western Atlantic Ocean	Warm, humid, and unstable	North America, east of the Rocky Mountains
Maritime polar, Pacific (mP)	North Pacific Ocean, near the Aleutian low-pressure area	Cool, moist, and unstable in the winter; cool and stable in the summer	Pacific Coast of North America
Maritime polar, Atlantic (mP)	North Atlantic Ocean off Newfoundland	Cool, moist, and unstable in the winter; cool, dry, and stable in the summer	Northeastern United States and eastern Canada
Superior (S)	Upper levels of troposphere in vicinity of subtropical high-pressure area	Hot, dry, and stable	Usually aloft, but occasionally appears at the surface in the southwestern and Great Plain States.

summer weather is less changeable than winter weather. Southern Hemisphere cyclones are usually better developed and travel more rapidly and in a more regular west-to-east direction, because of the relatively small size of the midlatitude land masses.

SIZE AND STRUCTURE

On a weather map cyclones are represented by a series of closed, oval-shaped isobars (Fig. 4–2). Although no two cyclones are exactly the same, they are usually egg-shaped, with the longer axis trending in a general southwest-northeast direction in the Northern Hemisphere. They generally range in size from 1000 to 1500 miles in diameter along their longer axis and 500 to 800 miles in width. The typical cyclone, then, covers several thousands of square miles in horizontal area. On occasion large storms may cover 1 million square miles or more, or about one-third the area of the United States. Pressure differences between the center and edge of the low usually range from 10 to 30 millibars (3.4 millibars = 1/10 inch of mercury). A large, intense cyclone may show pressure gradients of 40 millibars or more. On the whole, however, midlatitude cyclones have moderate pressure gradients and are nonviolent. They are extensive, weak low-pressure systems accompanied by clouds and perhaps precipitation, in contrast with such intensive, small storms as the hurricane or thunderstorm. Cyclones normally have a vertical development of 30,000–40,000 feet, extending to the top of the troposphere.

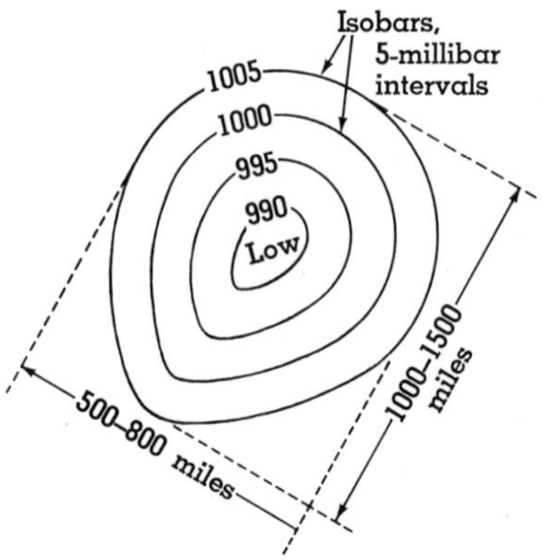

In a cyclonic storm, air masses of different properties converge but are separated from each other by surfaces of discontinuity, or fronts. For example, in central United States cold cP air is attracted southward by a passing low, and warm mT air from the Gulf is moved northward as a tongue of air bounded on its western, northern, and northeastern margins by cold air of polar origin. Within a cyclone, along the margins of the warm air, two types of fronts can be distinguished; these fronts are referred to as cold or warm, depending upon which of the air masses is the invading one. The over-all low is moving eastward, and the circulating air may be thought of as a large eddy or whirlpool being carried in a general west-to-east direction within a gigantic river of air represented by the belt of the westerlies. When a specific locality experiences an invasion of warm air, we say that a warm front has passed the area; similarly, a cold front is the boundary in front of incoming cold air (Fig. 4–3).

Warm front. As warm air moves northward and meets colder and denser air, it is wedged up in a gentle incline. This slope is very gradual— for every foot the warm air rises vertically, it moves horizontally 100 to 500 feet (1:100–500). The air cools adiabatically, and usually condensation and clouds occur. There will probably also be widespread rain, covering an area several hundreds of miles ahead of the surface front (Fig. 4–4).

Cold front. Here cold air is advancing rapidly and vigorously forces the warm air to rise as it is displaced. Cold fronts are four to five times steeper than warm fronts, and the most violent weather conditions within a cyclone develop along, or slightly ahead of, the surface front. If the cold

FIG. 4-2. Shape and dimensions of a model midlatitude cyclone.

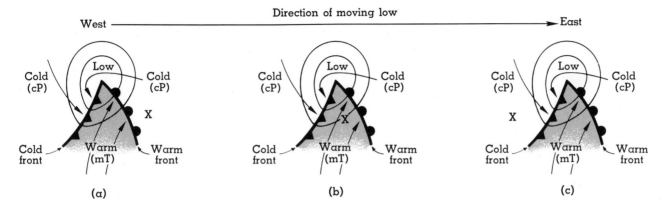

(a)   (b)   (c)

air advances rapidly, the air at the surface is slowed by friction and the air aloft may move ahead at a greater speed. When the cold air aloft pushes ahead of the surface front and overruns the warm air at the surface, highly unstable conditions may occur. If the warm air is very moist and contains a great deal of latent heat, severe thunderstorms and tornadoes may develop (Fig. 4–4).

WINDS AND PRECIPITATION

Because of Coriolis force, winds blow into a low in a counterclockwise direction in the Northern Hemisphere and clockwise in the Southern Hemisphere. Conversely, winds blow out of a high, or anticyclone, clockwise in the Northern Hemisphere and counterclockwise in the Southern Hemisphere. The winds in the southern sector of a low (Northern Hemi-

FIG. 4-3. Convergence of air masses in a cyclone. (a) An observer at point X is in the cold air mass but will experience an invasion of warm air shortly, as the low proceeds eastward. (b) The warm front has passed and warm air has replaced the colder air at the surface. (c) The cold front has passed over the observer; cold, dry air has displaced the warm, moist air. The low moves eastward at 20 to 40 miles per hour, and the sequence described would most likely take place over a 12- to 24-hour period.

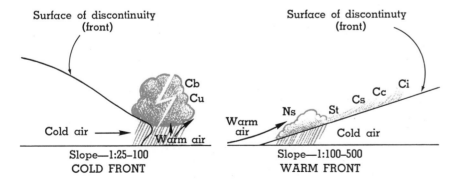

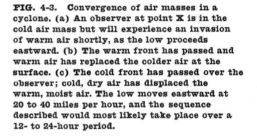

FIG. 4-4. Cross section of typical warm and cold fronts showing possible cloud types and precipitation areas.

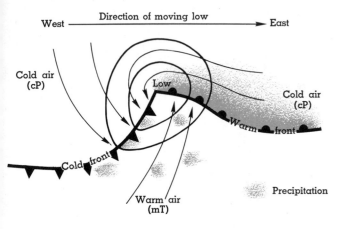

West ——— Direction of moving low ——→ East

Cold air (cP)

Low

Cold air (cP)

Warm front

Cold front

Warm air (mT)

Precipitation

FIG. 4-5. Wind directions and areas of possible precipitation in a model cyclone.

sphere) are southerly; northwesterly in the northern and western sectors; and easterly or northeasterly in the eastern sector, reflecting the pressure gradient and the deflection caused by the earth's rotation (Fig. 4–5).

Precipitation may occur in three areas within a cyclone (Fig. 4–5). *Warm-front precipitation* is associated with widespread clouds and steady rain or snow of long duration; it usually covers an extensive area of several thousand square miles. Drizzlelike rain lasting 12–24 hours, or prolonged and heavy snowfall in the winter, are usually the result of warm, moist air from the south wedged over cold, dense air along a warm front. Warm-front precipitation normally occurs to the east and northeast of the low center.

Cold-front precipitation is likely to occur along and slightly ahead of the cold front. The advancing cold air usually forces the warm air aloft with vigor, and this produces short, sharp, and often violent thunderstorms. The cold front may be an area of severe turbulence, with thunderstorms and squalls strung like beads on a string along the front. The area of precipitation is narrow, and the rainfall, although often torrential, is of short duration.

Finally, *nonfrontal* or *convergence precipitation* occurs in the southern area of the low in the warm-air sector. It is not directly associated with the fronts but results from the forced ascent of unstable air as part of the general convergence of air, or from surface heating. In any given low, precipitation will not necessarily occur in all three areas. Nor are the three areas always completely distinct from one another, since merging and modifications often take place.

SEQUENCE OF WEATHER IN A PASSING CYCLONE

The passing of a low brings about an interplay of air masses resulting in marked, and often abrupt, changes in humidity, temperature, precipitation, pressure, and wind direction and velocity. The classical sequence of weather in the passing of a well-developed, mature cyclone takes the following form, as illustrated in Fig. 4–6.

Diagram (a): Weather conditions are clear, with scattered high cirrus (Ci) clouds moving easterly in a well-defined pattern; surface winds are moderate from the east or southeast as a result of the low-pressure area to the west.

Diagram (b): The low has proceeded eastward during the course of several hours. The barometer has been falling gradually as the low center approaches, usually dropping at the rate of 2 or 3 millibars every 3 hours. A drop of 8 to 10 millibars in this period indicates a strong low with extreme conditions. The temperature has been gradually rising as the warm front approaches. The cold wedge is thinner, and cirrostratus (Cs) and cirrocumulus (Cc) clouds cover most of the sky. They have become noticeably lower and thicker during the preceding few hours.

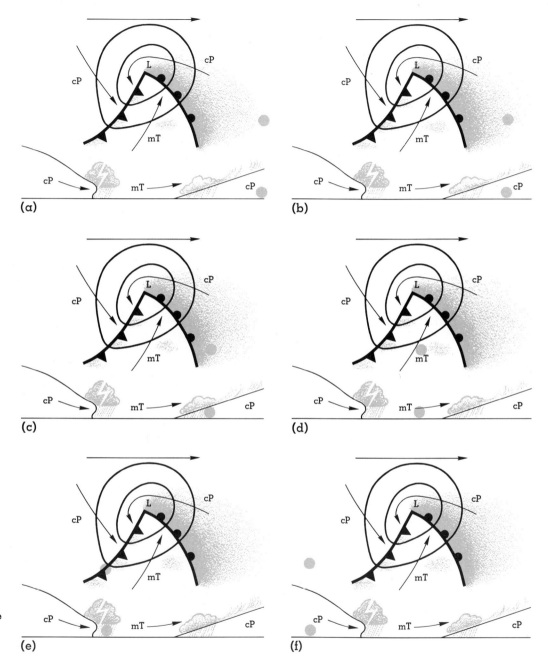

FIG. 4-6. Sequence
of weather in
a passing cyclone.
● Observer.

Diagram (c): A few hours later the clouds have become even lower and have gradually merged into widespread overcast of the stratus (St) and nimbostratus (Ns) types. Widespread steady rain may occur, or heavy snow if it is the winter season. The barometer is still falling and temperatures are gradually rising as the low center and warm front approach. Winds continue from a general southeasterly direction.

Diagram (d): The warm front has passed, and warm air (mT) has replaced the cP air. At the same time, the winds have shifted to the southwest, the temperature and specific humidity have increased rapidly, and the widespread overcast and drizzlelike rain have disappeared to the east. At present, the temperature and humidity are high—reaching their highest point in the passing of the low—and the barometer has dropped. Skies are partly cloudy, with possible scattered showers and southerly, warm, moist winds.

Diagram (e): The cold front is passing and turbulence may occur. Thunderstorms, squall lines, and tornadolike conditions are all possible. When the cold front passes, very rapid weather changes take place. Winds shift abruptly to the northwest, pressures rise rapidly, and temperatures

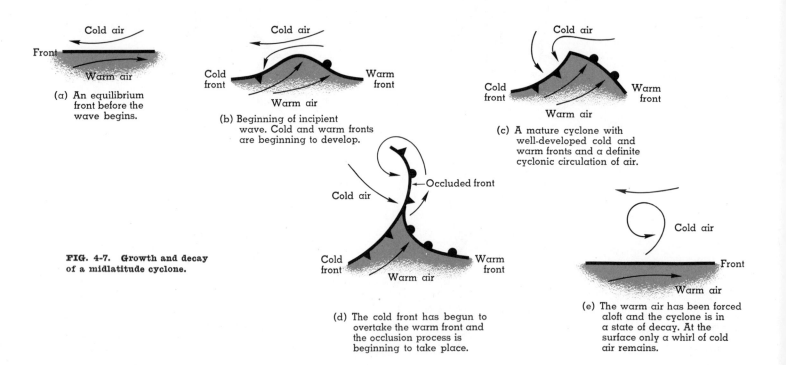

Cold air

Front

Warm air

(a) An equilibrium front before the wave begins.

Cold air

Cold front

Warm front

Warm air

(b) Beginning of incipient wave. Cold and warm fronts are beginning to develop.

Cold air

Cold front

Warm front

Warm air

(c) A mature cyclone with well-developed cold and warm fronts and a definite cyclonic circulation of air.

FIG. 4-7. Growth and decay of a midlatitude cyclone.

←Occluded front

Cold air

Cold front

Warm front

Warm air

(d) The cold front has begun to overtake the warm front and the occlusion process is beginning to take place.

Cold air

Front

Warm air

(e) The warm air has been forced aloft and the cyclone is in a state of decay. At the surface only a whirl of cold air remains.

nosedive. A drop of 20–30° in one hour is not uncommon. The incoming cP air is dry, humidity decreases abruptly, and the skies clear rapidly.

Diagram (f):   The low has passed and a high, or anticyclone, dominates the weather. Clear, cold, dry conditions prevail, and strong winds blow from the northwest.

It should be stressed that the full sequence of weather just described occurs only where a cyclone in the mature stage passes directly overhead. Where a low center passes to the north or south, the sequence of weather is modified; the surface fronts and the associated abrupt weather changes may not be experienced. There also are deviations from the classical sequence for a cyclone in its old-age period. Cyclones go through stages of growth and decay (Fig. 4–7). In the period of decay, the cold front moves faster than the warm front and eventually merges with it, displacing the warm air at the surface and forcing it aloft. This usually takes place in 3–5 days and is the beginning of the end of the cyclone.

According to a generally accepted theory developed by a group of Norwegian meteorologists, cyclones begin as a wave in the subpolar front separating semitropical and polar air. The wave causes warm air to extend poleward and replace cold air. The warm air is also wedged aloft over the cold air resisting its advance at the surface. The cold air at the rear of the wave then underruns the warm air and forces it up, and the beginnings of a warm and a cold front can be distinguished, as well as a counterclockwise circulation of air. In the mature stage, the fronts are well defined and the circulation of air reaches maximum development. The cloud system attains its greatest development, and precipitation areas are easily distinguished. Following the mature stage, the cold front begins to overtake the warm front and the surface tongue of warm air becomes occluded, or pinched out. Eventually the cold front completely overtakes the warm or occluded front and all the warm air is forced aloft. At the surface, a whirl of cold, homogeneous air erases the surface wave completely.

As previously mentioned, the life cycle of a cyclone is usually completed in a 3-to-5-day period, although individual cyclones may last much longer under ideal conditions. The cyclone's major source of energy lies in the contrast between the temperature and moisture conditions of two air masses. Latent heat of condensation is an additional source of energy, particularly if the warm air contains a great deal of moisture. Regions of the world where cold cP air and warm mT air can be brought together without their characteristics having been modified en route are ideal sites for the development of cyclones. Such sites are called *areas of frontogenesis.* The region south of the Greenland ice cap along the eastern coast of North America is an excellent example.

All parts of the world under the influence of the belt of the westerlies are affected by moving cyclones and anticyclones. However, low centers travel along certain paths more frequently than others. In the Northern Hemisphere during the winter season, the Icelandic, Mediterranean, and Aleutian lows are regions where cyclones develop and which also attract moving lows. Lows attract lows, and high air pressures repulse moving lows. Thus, the strong continental highs serve as a block forcing moving lows to detour to the south or north. The large land masses at midlatitudes in the Northern Hemisphere cause irregular cyclone paths, whereas in the Southern Hemisphere, in contrast, cyclone paths exhibit considerable west-to-east regularity (Fig. 4–8).

ANTICYCLONES

Anticyclones, or moving high-pressure centers, are usually 20–30 per cent larger than cyclones. They are characterized by clear, dry, and cool weather, descending currents, and diverging winds. They also move from west to east and are found chiefly in the belts of the westerlies, at latitudes 20° to 70° in both hemispheres. In the winter season, cold waves and very low temperatures are always associated with anticyclones. Bitter cold reflects the polar origin of the anticyclone and the very dry air that allows rapid terrestrial radiation to take place. In the summer, weak highs that are subtropical in origin may become stagnant over the continents, slowly moving over the central and southeastern parts of the country. Excessively high temperatures, or hot waves, are the result of the invasion of these highs, composed of warm, subtropical air.

FIG. 4-8. Highly generalized map of the most common storm tracks.

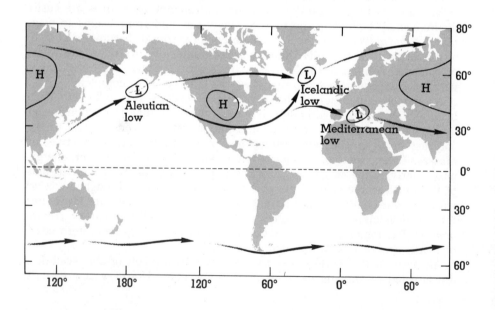

EARTH SCIENCE

During the winter season moving cyclones and anticyclones are the major control of weather conditions in the midlatitudes. The interplay of air masses results in sudden changes in temperature, humidity, pressure, wind direction, cloud cover, and precipitation. Cyclones and anticyclones are best developed in the winter and when the jet stream is wavy or irregular. The storm paths migrate, following the vertical ray of the sun, and dip equatorward as far as latitudes 20° to 25°. In the summer, the storm tracks move poleward to latitudes 50° or 60° Cyclones are poorly developed then and play a subordinate role in controlling weather conditions. The angle of the sun and length of day are dominant and produce less-changeable day-to-day weather.

VIOLENT STORMS

The hurricane, tornado, and thunderstorm, although not highly significant over a climatic region, are of importance meteorologically and, of course, are of the utmost concern locally. These atmospheric disturbances are among nature's most violent phenomena and take a high toll in human life and property damage.

THE HURRICANE

The hurricane is a tropical storm known by different names in different parts of the world. It is *typhoon* in the Western Pacific, *tropical cyclone* in India, *baguio* in the Philippine Islands and the South China Seas, and *willy-willy* in Australia.

Hurricanes resemble midlatitude cyclones insofar as both are areas of low air pressure, and winds blow into both in a counterclockwise direction in the Northern Hemisphere and clockwise in the Southern Hemisphere. Here the resemblance ends, for there are differences of size, structure, areas of occurrence, and season of frequency. The hurricane is 100 to 500 miles in diameter and is depicted on a weather map as nearly circular, with the isobars very closely packed and the pressure gradient exceptionally steep (Fig. 4–9). The lowest officially recorded sea-level pressures in the world have been found in the low centers of these storms; readings of 27 inches are not uncommon, and in one instance a sea-level reading of 26.35 inches (892 mbs) was recorded. Pressure drops of 0.6–0.8 inch per hour are not uncommon, and drops of over 1 inch in 30 minutes have been recorded. The steep pressure gradients result in winds of high velocity. A wind of 75 miles per hour or more is considered a hurricane wind. Velocities of over 100 miles per hour have been recorded, and it is estimated that winds of over twice that velocity have probably occurred.

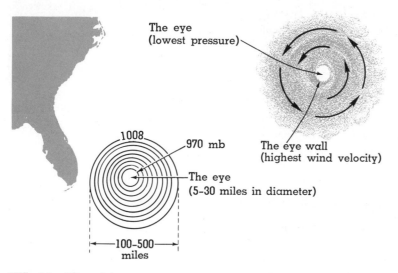

FIG. 4-9. Dimensions and structure of a model hurricane.

The eye
(lowest pressure)

The eye wall
(highest wind velocity)

1008

970 mb

The eye
(5–30 miles in diameter)

100–500
miles

FIG. 4-10. View of a hurricane from a weather satellite in orbit. Hurricane Betsy, transmitted from **Tiros VIII** in September, 1965, showing circular pattern of clouds and the eye of the storm. (Environmental Science Services Administration, **U.S.** Department of Commerce.)

The hurricane is a gigantic whirl of air with a calm central core, or *eye* (Fig. 4–10). The eye is 5–30 miles in diameter and is characterized by relatively clear skies, little or no rain, and moderate winds. Around the eye, clouds form spiraling bands known as the *eye wall*, and it is here that the storm's fury is concentrated in the form of strong winds and torrential rains. Hurricanes develop within a maritime tropical air mass, and well-defined fronts, with associated abrupt wind shifts, precipitation areas, and temperature differentials, are unknown. Rainfall and temperature are relatively equally distributed around the center. Recent data have shown that temperatures are slightly higher in the interior of the hurricane than on the outside, though the reason for this is not fully understood. The eye wall of the storm extends vertically to the base of the tropopause, reaching heights of 5–7 miles.

Origin and areas of occurrence. Hurricanes develop in the tropical latitudes but may on occasion invade the midlatitudes. They rarely form poleward of latitude 20° or nearer to the equator than 5°. The fact that they do not occur at or close to the equator probably stems from the weakness of the Coriolis force there and the consequent inability of the cyclonic whirl to develop. Hurricanes always form over warm water; evidence indicates that sea-surface temperature must be at least 79° F., and most storms develop when the surface temperature is 82° F. or higher. A hurricane weakens rapidly if it moves over land or cool water. Warm water is essential to its development and continued existence, for the air above warm water becomes very humid, and latent heat of condensation is the source of the hurricane's energy.

The storm begins as a gentle, westward-moving wave associated with scattered showers. We do not understand why some of these easterly waves remain relatively mild and others develop into violent cyclonic whirls. But once started, hurricanes are steered westward by the trade winds at speeds of 5–30 miles per hour. On occasion they slow down and stagnate over an area, causing heavy rains and flooding. As they travel, they are deflected by Coriolis force to the right in the Northern Hemisphere and to the left in the Southern Hemisphere, taking a parabolic path. The average life of a hurricane is 8–10 days, but if its path coincides with warm ocean currents, it may last longer and may move into latitudes of 45° or more. There have been storms of this type that have followed the course of the Gulf Stream and have moved northward along the eastern seaboard as far as New England. Normally the storm's passage over any specific locality takes 12 to 24 hours. Most hurricanes develop in late summer or in autumn, though some have been known to occur in every month of the year. In the Northern Hemisphere, the months of greatest frequency are August, September, and October. This concentration roughly coincides with the furthest poleward migration of the equatorial low and also with the season of highest sea-surface temperatures. In the Caribbean and Atlantic areas, the average number of hurricanes per year is 7, with a range of 1 to 11 per season. In the North Pacific, the average is 21 per year.

Hurricanes occur in six general areas (Fig. 4–11):

Caribbean Sea, Gulf of Mexico, and adjacent North Atlantic Ocean.
Pacific Ocean, west of Mexico and Central America.
Pacific Ocean, vicinity of the Philippine Islands and China Seas.
Bay of Bengal and Arabian Sea.
South Indian Ocean, east of Madagascar.
South Pacific, off the north and east coasts of Australia.

It should be noted that most hurricanes occur off the east coasts of continents. This reflects the pileup in these areas by the trade winds of warm surface water that has been heated by the tropical sun. As the surface water is blown westward, cool water upwells along the west coasts, decreasing the chances of hurricane development. Sea-surface temperatures may be 10–30° warmer off the east coasts of land masses than off the west coasts. The only notable exception to this observation is the area along the west coast of Mexico and Central America. A partial explanation is that most hurricanes occurring here actually originate in the Caribbean Sea and cross the very narrow land bridge of Central America or Mexico without being neutralized.

FIG. 4-11. Generalized map showing the regions of hurricane occurrence. Arrows indicate the principal tracks and direction of movement; note the parabolic paths.

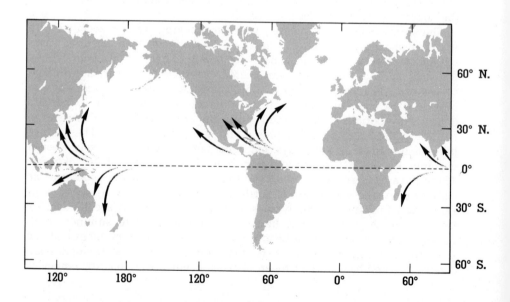

FIG. 4-12. Damaged homes at Virginia Beach, Virginia, the result of high seas and waves associated with the passing of a hurricane. (U.S. Department of Agriculture.)

Hurricanes do not form off the coast of South America—the only region of the world off the east coasts of continents between latitudes 5° and 20° where they are unknown. This anomaly results from the coastal outline of South America. The elbow of the continent (Cape Sao Roque) is south of the equator, and the warm surface waters moving westward in the Southern Hemisphere near the equator are deflected into the Northern Hemisphere's circulation. Thus the lack of warm surface water precludes the development of such storms in this region.

Effects of hurricanes. The disastrous effects of hurricanes in loss of life and property damage are well known. The damage is inflicted by high winds, torrential rains, and sea waves (Fig. 4–12). We have previously mentioned that velocities of over 100 miles per hour have been recorded and that hurricane winds of 200 miles per hour or more have probably occurred. Wind damage can be very severe, for the pressure on buildings and other structures increases proportionally to the square of the wind speed. It is not uncommon for hurricanes to bring 20–30 inches of rainfall to a given locality. In an extreme case, 96.5 inches of rain fell during the passing of one hurricane in Jamaica over a 4-day period. Think of this in contrast with the fact that the total average rainfall for an entire year for Chicago, Illinois, is 33 inches.

As a result of torrential rains, disastrous flooding frequently occurs. Strong winds create sea waves 10–15 feet in height. In low coastal areas, the sea has been known to advance inland to a distance of 50–60 miles, and the greatest damage in a hurricane is often the result of sea-wave flooding. Huge ocean swells moving faster than the storm may reach the coast 1 or 2 days ahead, while the storm itself is still 300–500 miles out to sea. Recently, observations of the dimensions and direction of these waves as they reach the coast have been used to predict the time and place of the storm's arrival and its general intensity.

THE TORNADO

Although very small, the tornado is the most violent and destructive atmospheric disturbance in nature. Tornadoes are referred to in some areas as *cyclones* or *twisters*. They are funnel-shaped clouds that average 300–400 yards in diameter but may be as small as 9 feet or as large as 1 mile or more (Fig. 4–13). They are of relatively short duration and travel an average distance of 16 miles. In one instance, however, a tornado was known to travel over 290 miles in 7½ hours. The average total area covered is about 3 square miles, and the time normally required to pass over a given locality about 30 seconds.

Tornadoes are barometric depressions and in this sense resemble midlatitude cyclones and hurricanes. They are much smaller, however, and their pressure gradient is several hundred times as great. No accurate data have been recorded within a tornado, since no weather station can withstand the violent conditions. But detailed analysis of damage indicates that pressure drops of 3–5 inches or more take place over a distance of a few hundred yards. This extreme pressure gradient causes winds close to the funnel to reach velocities of 200–500 miles per hour or more. These are the strongest surface winds in nature. Fortunately, their paths cover only a few hundred yards. At the center there is a very violent updraft, which ordinarily attains a velocity of several hundred miles per hour. The tornado might be described as nature's vacuum cleaner.

Mode of occurrence. Tornadoes have probably occurred on every continent at one time or another, but they are primarily a phenomenon of North America. In recent years, about 200 per year have been recorded in the United States. More have been reported of late than in past years, and this is no doubt the result of the development of an efficient network of radar installations. Tornado echoes appear on radarscopes as small button-

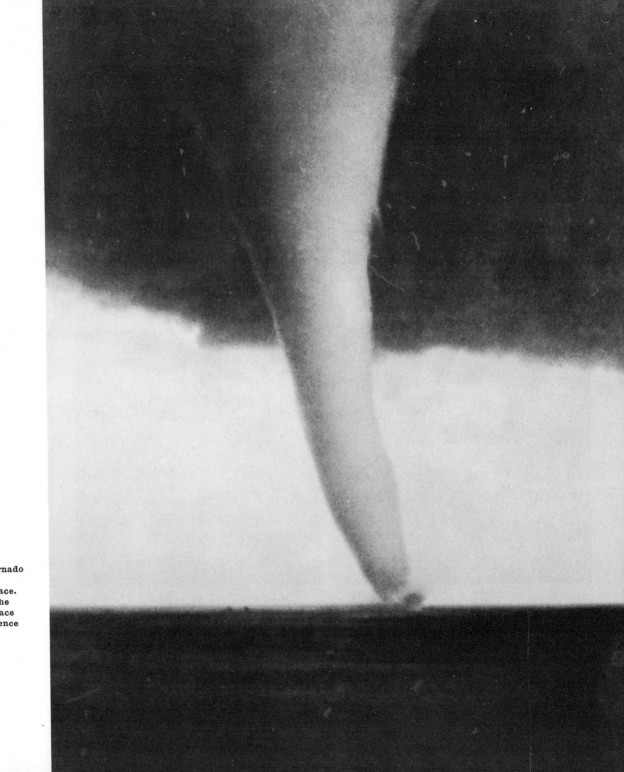

FIG. 4-13. A well-defined tornado funnel. Note the heavy, dark clouds with ragged undersurface. The funnel may move along the ground or skip above the surface at times. (Environmental Science Services Administration, U.S. Department of Commerce.)

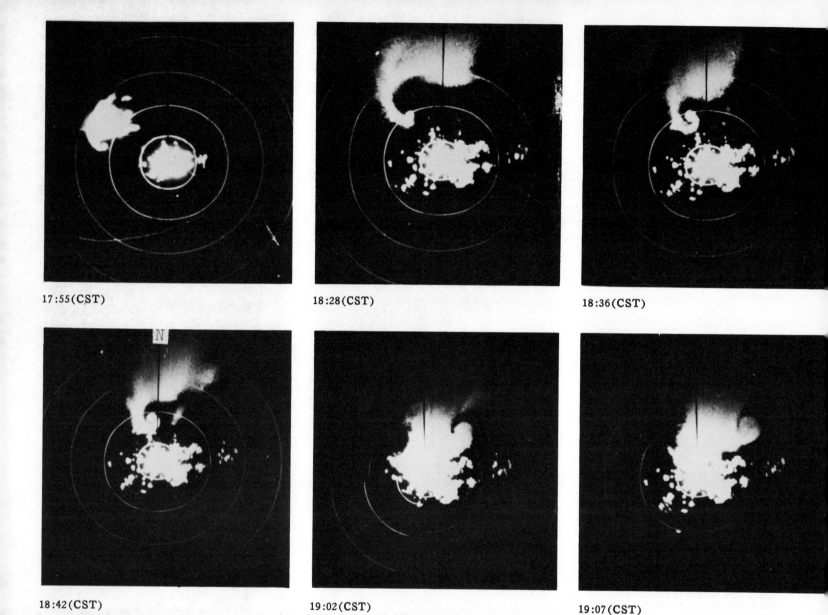

17:55(CST)

18:28(CST)

18:36(CST)

18:42(CST)

19:02(CST)

19:07(CST)

FIG. 4-14.   Radar view of successive phases of a tornado; Meridan, Kansas, May, 1960.
The buttonhook-shaped funnel is clearly seen. Speed of movement can be inferred
from the time lapse of little more than 1 hour between the first view and the last.
(Environmental Science Services Administration, U.S. Department of Commerce.)

96

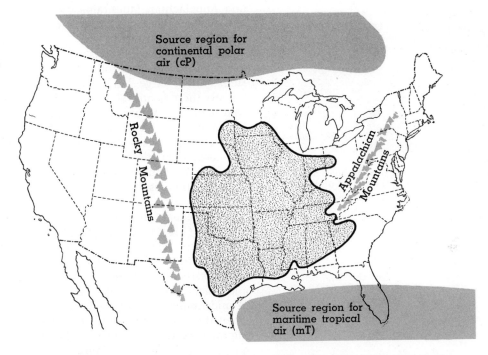

Source region for
continental polar
air (cP)

Rocky Mountains

Appalachian Mountains

Source region for
maritime tropical
air (mT)

hook-shaped objects and can be detected over 200 miles away (Fig. 4–14). Although there have been tornadoes in every state, the greatest concentration is in the central and midwestern parts of the country (Fig. 4–15). The greatest frequency per unit area is in Iowa, which averages 2.8 tornadoes per 10,000 square miles per year. Kansas, Oklahoma, Arkansas, and Texas also have high frequencies.

The concentration of tornadoes in central United States is readily understood in terms of the physical location and topography of the North American continent. Tornadoes develop along the cold front of a well-developed cyclone, where air masses of considerably different characteristics come together and cold, dry air (cP) overruns warm and very moist air (mT). In the early spring, polar air over the ice- or snow-covered regions of Canada and the Arctic is as cold and dry as it is in midwinter, while air over the Gulf of Mexico is about as warm and humid as it is in the summer. The size and location of the continent at midlatitudes thus allows for an exceptionally well-developed cP air mass adjacent to a large body of tropical water over which a humid and warm mT air mass may form. Furthermore, the lack of any major topographical obstacles or large water bodies between the Rocky Mountains on the west and the Great Lakes

FIG. 4-15. Generalized map showing the areas where tornadoes occur most frequently; based on over 8000 tornadoes recorded since 1916.

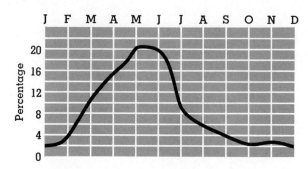

FIG. 4-16. Percentage of tornadoes occurring each month of the year in the United States.

and Appalachian Mountains on the east makes it possible for a passing low to move these air masses together in an unmodified or only slightly modified state.

Because such conditions as these do not exist elsewhere in the world, tornadoes rarely occur outside of central North America. In the Southern Hemisphere, large land masses are lacking at midlatitudes and consequently there is no well-defined cP air mass; and the cP air of the Eurasian continent is separated from mT air by an extensive east-west mountain region having the highest and largest mountains in the world. Any air mass that crosses such imposing highlands is neutralized as it ascends and cools adiabatically and descends and is heated by compression. Southeast Asia is dominated by the monsoonal system of winds, and passing lows are thus less important there as a control of weather than in North America.

As previously mentioned, tornadoes are most frequent in the spring of the year (Fig. 4-16). Although there have been some in every month, about 68 per cent occur from March to June. The reason for this seasonal concentration is that at this time the cP air still retains most of its winter characteristics, while the mT air, owing to the latitude of its source region and much earlier summer, has characteristics similar to those it possesses in midsummer. The spring season, therefore, is the only time of the year when the maximum contrast between air masses exists. The tornado belt migrates latitudinally during the spring, so that tornadoes are most frequent in February and March in the South Central States and in May and June in the North Central States.

Tornadoes usually occur slightly ahead of the cold front and are steered by the southwest winds in the southern sector of a midlatitude cyclone (Fig. 4-5). For that reason over two-thirds of the tornadoes recorded have traveled in a general southwest-to-northeast direction. They meander in their course owing to local topographical and pressure conditions, usually moving at a speed of 20–45 miles per hour but on occasion slowing down or else accelerating to 65 miles per hour or more for short distances. The rapidly moving cold air pushes ahead of the surface cold front slowed by friction and overruns and traps warm, humid air below. The cold air aloft may extend 50–100 miles ahead of the surface cold front at heights of 1000–3000 feet above the surface. Under such conditions, there is a steep lapse rate that reaches the mechanical-instability stage (19° per 1000 feet), and the air automatically overturns. It has been observed that when tornadoes develop, the jet stream is wavy and well developed. Thus it is thought likely that the jet stream plays a part in tornado development.

Effects of tornadoes. Although tornadoes are small and relatively infrequent, they may bring great destruction. The violence of a tornado is

FIG. 4-17. Tornado damage. The explosive effects of great local differences in air pressure, violent winds, and hail can result in tremendous property damage and loss of life. Note the selective nature of property damage. (Environmental Science Services Administration, U.S. Department of Commerce.)

due to the strong winds, the violent updraft, and the explosive effect of the great pressure differential—the abrupt reduction of pressure outside a building can make it literally explode (Fig. 4–17). Violent rain and hail may also inflict severe damage.

THUNDERSTORMS

Thunderstorms are local disturbances characterized by violent turbulence; fast, sharp rain, and sometimes hail; gusty squall winds; and lightning and thunder. They develop in warm, moist unstable air and require the release of large amounts of latent heat and the availability of ice crystals to serve as nuclei for raindrop formation. Under suitable conditions, a cumulus (Cu) cloud develops into a cumulonimbus (Cb) cloud, or thunderhead, often extending to the tropopause. Precipitation is usually heavy but of short duration. Hail may develop if the cumulonimbus cloud is exceptionally intense. These storms cover only a few square miles and usually last about 30 minutes. In that brief time, ½–2 inches of rain may fall and cause local flooding. The boundaries of the storm are well defined; rainfall may be heavy in one locality, while no rain occurs a few blocks away. The cumulonimbus cloud is made up of several independent cells of air circulation and, within each cell, updrafts and downdrafts that may reach velocities of 70 miles per hour or more and thus present a serious hazard to flying aircraft.

Thunderstorms are often preceded by squall winds. These are gusts of cold air brought down from aloft by downdrafts in the storm. The colder and denser air spreads out in the advance of the storm underrunning the warmer air. Squall winds are usually gusty and may attain velocities high enough to cause local damage.

Lightning is an electrical discharge that heats the air along its course as much as 18,000° F. A tremendous expansion of the air results, and a pressure wave is thrown outward, producing thunder. Lightning may occur within a cloud; from one cloud to another; or, less frequently, from a cloud to the surface of the earth. The generation of electricity within a thunderstorm is not fully understood, but it is thought to be the result of the breaking up of large raindrops. Raindrops initially carry equal amounts of positive and negative electrical charge and are consequently neutral. As they fall, positive or negative charges migrate to the head of the raindrop and the opposite charges move to the tail. This rearrangement may be due to the attraction of the earth's surface. If the earth is negative in relation to the raindrop, positive charges will migrate to the head of the drop, and if the earth is positive, they will move to the tail, since unlike charges attract and like charges repulse. As the drops grow larger, they may reach a size at which the limit of their stability, or cohesion, is reached, and they break up. The splitting of the drops then produces concentrations of negative and positive charges in different parts of the cloud. When the accumulation of opposite charges becomes great enough, lightning results. It is possible to see lightning when no rain is visible, but one may then assume that the rain is taking place in a cloud. Since lightning occurs only in turbulent clouds of the cumulonimbus type, it is most frequent during the warm season and in warm regions.

Thunderstorms are grouped into two major types: (1) air-mass storms and (2) frontal storms. Air-mass storms take place within a warm, humid air mass as the unstable air is forced aloft. This group includes local heat thunderstorms, in which air heated at the surface rises in the form of a convectional current; orographic thunderstorms, due to the forced ascent of horizontally moving air over highlands; and thunderstorms resulting from advection of warm air near the surface or the overrunning of cold air aloft. These storms are local, occurring as scattered, isolated disturbances.

Frontal thunderstorms are the result of the interaction of two air masses as a front passes a given locality. Cold-front storms occur as cold air underruns warm air by forcing it aloft; they develop along or ahead of the cold front, or wind-shift line. There are usually several such storms arranged along, or 50–300 miles in advance of, the front. They are usually more severe than air-mass thunderstorms and are always followed by cooler and drier weather as the cold air mass displaces the warmer air mass at the surface. Thunderstorms associated with a warm front are not nearly so common as those of the cold-front variety. They occur only when the warm air is at very high temperature and humidity and very unstable. The warm air is then forced to ascend over the cold air in a gradually inclined slope and, if a great deal of latent heat is released when condensation occurs, the warm air will rise rapidly and steeply, producing a cumulonimbus cloud.

Geographic distribution of thunderstorms. It has been estimated that about 44,000 thunderstorms occur daily over the earth, and an average of 1800 at any given time. Since thunderstorms are associated with moist, warm, unstable air, they are most frequent in the humid tropics. Some tropical areas experience thunderstorms 200 or more days of the year. They rarely occur in midlatitudes in the winter season, and they are virtually absent from polar regions. In the United States, more thunderstorm days occur along the Gulf of Mexico, adjacent to the source region of mT air, than anywhere else. In general, the number of thunderstorm days decreases inland and poleward (Fig. 4–18). Some areas along the Gulf average 70 or more days in which thunderstorms occur, and the region around Tampa, Florida, averages 94 days, the highest in the United

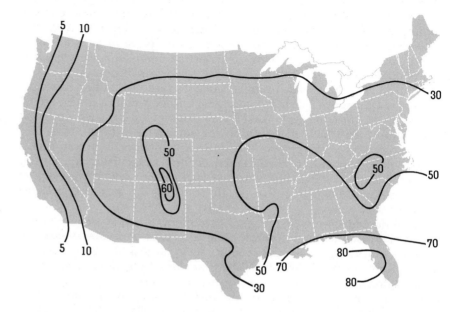

FIG. 4-18. Average annual number of days in which thunderstorms occur in the United States.

States. Another region of frequent thunderstorm activity is along the southern Rocky Mountains, where air rises orographically. The Pacific Coast has fewer thunderstorm days than any other section of the United States, owing to the influence of stable mP air during the summer season.

SUGGESTED REFERENCES _____

Battan, L. J.: *The Nature of Violent Storms*, Doubleday & Company, Inc., Garden City, N.Y., 1961.

Byers, H. R.: *Thunderstorm Electricity*, University of Chicago Press, Chicago, 1953.

Dunn, G. E., and B. I. Miller: *Atlantic Hurricanes*, Louisiana State University Press, Baton Rouge, La., 1960.

Flora, S. D.: *Tornadoes of the United States*, University of Oklahoma Press, Norman, Okla., 1953.

Kimble, G.: *Our American Weather*, McGraw-Hill Book Company, Inc., New York, 1955.

Petterssen, S.: *Weather Analysis and Forecastings*, McGraw-Hill Book Company, New York, 1956.

Reiter, E.: *Jet Stream Meteorology*, University of Chicago Press, Chicago, 1963.

Riehl, H.: *Tropical Meteorology*, McGraw-Hill Book Company, Inc., New York, 1954.

Tannehill, I. R.: *Hurricanes: Their Nature and History*, Princeton University Press, Princeton, N.J., 1944.

CHAPTER 5

THE OCEANS

5

The absorption, transfer, and release of heat by the oceans have a substantial effect on weather and climate in many regions of the earth. To complement our consideration of the earth's atmospheric phenomena we turn now to a study of that part of the hydrosphere (the water of the earth) that exists in the oceans. In addition to their importance as a control of climate, the oceans provide us with much information about the earth's history, and they are a great and still largely untapped potential source of minerals and food.

The oceans are a vast reservoir of elements vital to man. Comprising about 98 per cent of the water of the hydrosphere, they are the source of a large percentage of atmospheric moisture and the ultimate destination of most of the surface and subsurface water of the continents. Many materials, including dissolved gases from the atmosphere and solids eroded from the continents, are distributed over the globe by ocean currents. Some of the gases return to the atmosphere or are used in biological processes; some mineral substances are retained in solution; and some solid particles are deposited as sediment. Sediment transported by currents has been found deposited in rock bodies with characteristics that indicate they are millions of years old. Ocean waters in low latitudes absorb heat from the sun, transport it by currents, and release it elsewhere, with great impact on both climates and biological activities.

For thousands of years, salt has been produced by evaporating sea water, and we now extract economically many tons each year of magnesium (a light-weight metal) and bromine (used in the chemical industry). Profitable production of manganese, essential in steel making, is expected in the near future. Of the hundreds of other elements and compounds dissolved in the sea or existing as deposits on the ocean bottom, the list of those being extracted in large quantities will constantly grow as deposits on land become exhausted.

We tend to take the oceans for granted as a source of food, specifically fish and other forms grouped in common parlance as "sea food"; but intensive oceanographic studies have been and are still being made with a view to increasing the yield by proper management. Soon man will engage in "farming" the sea in some localities by controlling nutrients, weeds, and predators, as he does on land. Further, as the pressures of population growth and the demand for augmented food supplies in many underdeveloped areas of the world become greater and greater, man will overcome his habits and prejudices to the point of harvesting from the sea food crops not now generally found acceptable.

PROPERTIES OF OCEAN WATER

PHYSICAL ASPECTS

Perhaps the most important single factor affecting the life processes that go on in the ocean, and one which exercises some degree of control on chemical and physical processes as well, is temperature, which is the direct result of heat flow in the near-surface materials of the earth. Some heat comes from within the earth, but the amount is insignificant compared with that from the sun. Solar heat emission is such that at our distance of 93 million miles from the sun, heat energy reaches the earth at the rate of 2 gram-calories per square centimeter per minute. In more familiar terms, the energy received by one square foot of surface is sufficient to heat a cup of water from 70° F. to the boiling point in about 20 minutes. This rate has apparently changed little in over 3 billion years. Not all the radiant solar energy approaching the earth reaches its surface, as discussed in Chapter 2.

The oceans are cooled primarily by heat loss to the atmosphere. In polar regions heat loss is almost perpetual; it occurs when the atmosphere is below 32° F., as heat moves from the water to the colder air even through intervening ice and snow. In middle and high latitudes during the winter months cold winds blowing from the continental semipermanent high-pressure areas toward the oceanic lows chill the surface water of the oceans near the shore. Wherever evaporation occurs, the water of the oceans is cooled. In addition, oceanic cooling from the last glaciation appears to be residual in some deep, cold waters.

CHEMICAL ASPECTS

Both solids and gases are dissolved in the oceans, and most play active roles in one or more of the organic or inorganic processes occurring there. The distribution and cyclic and noncyclic gain and loss of dissolved materials are matters of intense study by oceanographers, but as yet we cannot consider ourselves really knowledgeable about many aspects of the subject.

The *salinity* of the ocean refers to the content of dissolved salts. In the open sea it is about 3.5 per cent by weight but may be more or less locally depending on the effects of evaporation, precipitation, inflow of surface streams, and other influences. Although salinity varies from place to place, the relative amounts of the solid constituents are quite uniform; the major materials are shown in Table 5–1. In all, about 50 elements have been identified in sea water, and it is believed that all the known

TABLE 5-1 MAJOR DISSOLVED MATERIALS IN SEA WATER

MATERIAL	PER CENT OF TOTAL	PER CENT OF SOLIDS
Chlorine (Cl)	1.8980	55.05
Bromine (Br)	0.0065	0.19
Sulfate (SO_4)	0.2649	7.68
Bicarbonate (HCO_3)	0.0140	0.41
Fluorine (F)	0.0001	
Boric acid (H_3BO_3)	0.0026	0.07
Magnesium (Mg)	0.1272	3.69
Calcium (Ca)	0.0400	1.16
Strontium (Sr)	0.0008	0.03
Potassium (K)	0.0380	1.10
Sodium (Na)	1.0556	30.61
Total	3.4477	99.99

natural elements are present, although many are not in sufficient concentrations to be detected by existing methods of analysis.

The salinity represents a balance between the supply and the loss of both dissolved materials and water, which exist together in a complex assemblage. Salts are derived from rock weathering on the continents and, to a lesser degree, from particles of terrestrial and cosmic origin that fall from the atmosphere into the oceans and release materials after alteration there. Volcanic, eolian, and meteoritic dust dominate in the latter category.

Certain materials are selectively withdrawn by inorganic processes of sediment deposition and by life processes. Salt deposits, gypsum, and chemically deposited limestones are examples of sediment deposited by inorganic processes. Extraction by plants and animals for incorporation in skeletal materials leaves the occasional fossil in marine sediment, and, at the other extreme, the highly concentrated organic deposits of the coral reef. Adjacent to continents streams affect salinity by their contribution: In an arid climate highly saline streams and high evaporation rates may establish locally higher-than-average ocean salinity, and adjacent to a continent with abundant rainfall sea water may have less-than-average salinity.

The addition and extraction of water also affect salinity. In the doldrums, high rates and amounts of convectional precipitation reduce salinity appreciably in the near-surface waters diluted by rain. In contrast, water loss by evaporation in the belt of the subtropical high establishes belts of higher-than-average salinity. To a lesser degree polar freezing withdraws water, by forming ice and leaving the sea water enriched with the residual salts. These variations are latitudinal, but local variations may also exist. Fresh-water streams entering the sea may flow as a low specific gravity layer floating on top of the salt water for many miles out to sea, finally becoming mixed by diffusion or turbulence. Floe ice, pack ice, or shelf ice may melt, forming a layer of fresh water floating above normal sea water. Mixing will result in a reduced salinity of the sea water.

In the present, the oceans are a source of mineral substances and petroleum that are extracted by man; in the geologic past, dissolved salts were the sources of sedimentary rocks. Commercial deposits of a variety of minerals consist of beds of chemical sedimentary rocks of marine origin. Many kinds of dissolved materials are essential to life. Calcite and calcium phosphate are extracted for building internal and external skeletons of plants and animals, and other substances are used in smaller amounts for the growth of tissue. Of the elements withdrawn by organisms, some are returned to circulation by excretion or by death and decay of individuals; some are withheld in body tissues; and part may be withdrawn at one level of the ocean and returned at another, as by swimming organisms that feed at the surface in the daylight hours and sink to the bottom at night.

Most elements are used near the surface, and they may be returned there in any of three ways. *Upwelling* of deep waters takes place in areas where winds blowing off continents move surface water seaward, to be replaced by water from the bottom. The west margins of continents in low latitudes are the most likely places for upwelling. *Tidal currents* may bring deep water to the surface where the bottom is uneven or in narrow straits. Turbidity may also be increased, with the result of discouraging life. Vertical *convection currents* bring nutrients to the surface, especially in temperate zones in the winter.

Reverse motion may take nutrients down the sloping ocean bottom so that they are lost from circulation. Some substances are buried in the bottom sediment as skeletal material, organic matter, in trapped water, or adsorbed on clay particles and are then kept out of the cycle until the sediment is returned to the surface by crustal deformation and is thus exposed to the weathering process. The site of withdrawal becomes significant in this aspect of the cycle, for materials deposited in shallow water are likely to be returned to a subaerial environment, whereas deep-sea sediments are rarely exposed again. Phosphorus, nitrogen, and silicon are elements in great demand; of the three, phosphorus is most likely to become immobilized in deep-sea sediment and essentially permanently lost. This is believed by some oceanographers to be the limiting factor for the oceanic population.

Dissolved gases essential to biologic processes in sea water are oxygen and carbon dioxide. Their source, transportation, and destination are due

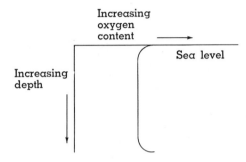

FIG. 5-1. Depth distribution
of oxygen in ocean water (diagrammatic).

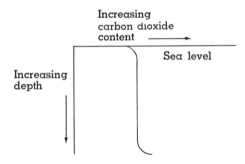

FIG. 5-2. Depth distribution of carbon
dioxide in ocean water (diagrammatic).

partly to physical and chemical factors and partly to biological ones, so the cycle is extremely complex and has not yet been fully clarified in all aspects.

Oxygen is not required in large amounts for plants, and in the open oceans it is generally present in adequate quantity. It crosses the atmosphere-water interface from the atmosphere at low temperatures and in the reverse direction where water and air are warm. Movement is thus related to latitude in equatorial and polar regions and to the season in midlatitudes. Oxygen is produced by plants in shallow water. Some is used by animals, some in oxidation of organic matter, and some is transported by currents to the deeper waters. Distribution with depth is shown diagrammatically in Fig. 5–1.

Carbon dioxide is essential to all life processes, since it is required for plants, which are the beginning of the food chain. It is used in chemical combination in the production of tissue and is involved in the secretion of skeletal matter composed of several minerals—two forms of calcium carbonate and, less commonly, magnesium carbonate and strontium carbonate. Many sources can be identified: As with oxygen, there is transfer from and to the atmosphere. Some originates in subaqueous oxidation of organic matter and some comes from animal metabolic processes. Substantial amounts are emitted volcanically, some is released during metamorphic processes, and man releases significant quantities by burning wood, coal, petroleum, natural gas, and other fuels. In solution in water, carbon dioxide forms carbonic acid (H_2CO_3) and may be involved in the solution of calcite. Some is used in chemical weathering of silicate minerals, but the amount thus withdrawn is small in the marine environment. The distribution of carbon dioxide in sea water with respect to depth is shown diagrammatically in Fig. 5–2; there may be local variations in the curve shown as well as in that of Fig. 5–1.

OCEANIC LIFE

Most organisms live in the *photic zone*, the layer of water in which a substantial amount of radiant energy is available, but some exist in the dark, cold water of the deepest parts of the oceans. Most organic matter is produced in the well-illuminated upper 30 feet of the photic zone, and some organisms pass their entire life in this depth range.

The organisms of the oceans are grouped in three categories, based on their place and mode of life. The term *plankton* refers to plants and animals that float and drift, unable to direct their motion. Most of them live in the photic zone. The force of gravity would cause them to sink to the bottom, but several devices allow them to remain in the photic zone through all or most of their lives. Some secrete skeletons which would increase their specific gravity to a value greater than that of sea water and would then sink. Certain forms are much flattened, which gives them such a low settling velocity that their entire life cycle may be completed without their sinking below the photic zone. Others secrete globules of fat or extract and store bubbles of gas in their tissues, both of which give a flotation effect.

Animals that can swim are called *nekton*. Included in this group are fish, some mammals, and many of the larger invertebrates such as squid, octopi, and jellyfish. Many of the important food products of the oceans are swimmers.

The third category of marine life is the *benthos*, the bottom dwellers, both plant and animal, ranging in size from microscopic to large. Burrowers, crawlers, forms which swim about but remain close to the bottom, and plants and animals attached to the bottom are included in this group. Significant food resources for man among the benthonic forms are oysters, scallops, shrimp, and prawns.

Microscopic plant plankton is the basic food of the ocean. Where there are light for photosynthesis, dissolved nutrient salts and carbon dioxide, and suitable temperature conditions, there we find plankton and all the forms that depend on them for food.

CONFIGURATION OF THE OCEAN BOTTOM

When we discuss the crust of the earth (Chapter 7), we shall find that there are differences between the continents and the ocean basins. The continental crust is thicker, comprising two layers (sial and sima) under the outer layer of sedimentary material whereas the ocean-basin crust has sediment and sima but no sial. Continental surfaces are higher than oceanic areas. Because subaerial processes differ from subaqueous processes, details of surface sculpture in the two environments differ substantially. Studies of ocean-bottom topography have revealed some of these differences, and our knowledge of the configuration of the ocean bottom is steadily expanding. But much more study will be required before we are really familiar with the shape of the ocean bottom, and still more before we know the materials below the uppermost few inches of sediment.

The ocean surface is about 72 per cent of the surface area of the earth, whereas the ocean basins together amount to about 65 per cent of the earth's surface. Thus the volume of water is slightly in excess of that required to just fill the basins; the water spills over and covers the continent surfaces around their margins to some extent. Continental heights above sea level range up to about 29,000 feet, in comparison to maximum oceanic depths of about 37,000 feet below sea level. Still greater contrast appears when average land surface above sea level—2700 feet—and average water depth—12,400 feet—are compared. Ocean basins are thus not only about twice as large in area, but over four times as deep as the height of the continental masses above sea level.

FEATURES MARGINAL TO CONTINENTS

Features of the subaqueous surface may be logically described by beginning at the shore of the ocean. The first element is the *continental shelf*, which is essentially the land surface of the continent continued in the seaward direction. The slope is of the order of 20 feet per mile over most of the shelf but varies from place to place (Fig. 5–3). The boundary is at the point of a distinct increase in slope that occurs at a surprisingly consistent depth of 400 to 600 feet. The width of the shelf ranges from essentially zero in a few places to as much as 200 miles. A relatively thick layer of recently deposited sediment derived from the continent is characteristic of the shelf. The photic zone makes up a comparatively large part of the water, and nutrients derived from subaerial rock weathering are plentiful, so life is abundant in this area.

The *continental slope* drops off into the deeper waters of the oceans from the edge of the shelf. The bottom gradient exceeds 1 to 40. Although this is greater than that of the shelf, it is still not large, as it amounts to less than 4° nearly everywhere. The lower limit of the continental slope is the point where the slope decreases at the top of the continental rise, at a depth between about 4500 feet and 10,000 feet. Sediment deposition is slower and its thickness less than that on the shelf, and both plant and animal life are much less abundant. Although the gradient is low, there is substantial evidence to indicate that much uncemented sediment on the continental slope is in an unstable condition and may be dislodged from place to move downslope into the deeper parts of the ocean basins.

The *continental rise* is a more gently sloping surface beyond the continental slope. Its gradient is less than 1 to 1000.

Abyssal plains are found in some places. They are extremely flat areas and occur at depths greater than the oceanic average. The Argentine Plain off the coast of Argentina has less than 10 feet of relief in a distance of over 800 miles. Sediment appears to be thick on the abyssal plains—for example, up to 2500 meters thick on the Argentine Plain. The sediment source is very likely turbidity currents (submarine currents containing a high percentage of bottom sediment and flowing along the ocean bottom) from the continental slopes.

POSITIVE FEATURES OF THE OCEAN FLOOR

Large, gently sloping positive areas are called *oceanic rises*. The highest point may or may not be above sea level. The volcanic peaks of the

FIG. 5-3. The gentle slope of the continental shelf shown in a view of the near-shore bottom at low tide. Shimaiko, Aioi City, Japan. (Japan National Tourist Organization, New York.)

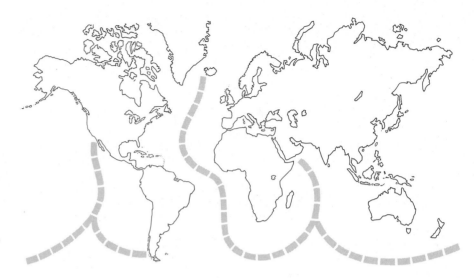

FIG. 5-4. Sketch map showing midoceanic ridges in dashed lines.

Hawaiian Islands are on an oceanic rise about 600 by 1900 miles in dimension. *Seamounts* are comparatively isolated single peaks or groups of peaks which may be entirely submerged or may be partly visible as islands. *Guyots* are flat-topped seamounts. A few are known in the Atlantic Ocean and several hundred have been found in the Pacific. Both vertical and horizontal dimensions have a wide range, the largest known guyot having an essentially flat top 35 miles in maximum dimension. The tops are 3000 to 5100 feet below sea level in the Pacific. The origin of the smooth tops is as yet unknown. Truncation by subaerial erosion has been considered as a possible answer, but some problems are unsolved by this hypothesis. The broad, flat surfaces are difficult to explain by wave erosion. On an extensive wave-eroded terrace near sea level the available energy of the waves moving across is rapidly reduced, and abrasion platforms the size of surfaces observed on guyots are difficult to account for within the range of time available for their formation. If subaerial or wave erosion was in fact involved, the depths and the pattern of distribution suggest that subsidence of the ocean bottom rather than a fluctuating sea level controlled the erosion. On at least one guyot, reef limestone several hundred feet thick overlies a basalt substratum.

The largest positive subsea features are the *midocean ridges*, which make up a continuous system of mountains. Those in the Atlantic and Indian Oceans lie along the axes of the oceans. In the Pacific Ocean the position is eccentric (Fig. 5–4). The mountains are of the scale of the cordillera of the North American continent. The topographic profile (Fig. 5–5) shows diagrammatically the central valley or cleft flanked by high mountains. In some places extrusive vulcanism has occurred along the Mid-Atlantic Ridge. The ridges are basaltic igneous rock, bare or covered by a thin layer of sediment. According to one hypothesis, the midocean ridges are zones where the crust of the earth is being torn apart, material is coming to the surface from the mantle, and the crust is thus being lengthened.

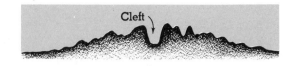

FIG. 5-5. Topographic profile of midoceanic ridge showing cleft.

Coral reefs are bodies of limestone deposited primarily as the result of life processes. The core of the deposit is essentially plant and animal skeletal material in place flanked by fragmental debris that originated in the reef. Horizontal dimensions range from a few hundred feet to several hundred miles. As seen in map view reefs can be divided into three types that appear to be stages in a continuously developing sequence. A *fringing reef* is built up along a shore and is the only structure between a landmass and the ocean. A *barrier reef* is some distance offshore, and a lagoon lies between it and the landmass. An *atoll* is a body of reef material that encloses a lagoon without an interior island or other landmass. Present distribution of growing reefs is from about latitude 25° south to latitude 20° north, somewhat modified by the temperature patterns of the oceans. Modern reefs require temperatures in excess of 60° F. for maintenance and growth, and optimum temperatures lie between 77° F. and 86° F.

Coral reefs begin as living, growing communities. Over half of the organic sediment in modern coral reefs is of plant origin (algae); corals of various kinds contribute the next largest amount, and many other groups contribute material, essentially all of which is calcium carbonate. The growth of a coral reef is not uniform, the rate being greatest where the environment is most favorable for growth of the organisms. Primary nutrients needed are carbon dioxide and dissolved salts. Carbon dioxide comes from the atmosphere and is therefore available in greatest quantities where wave action and turbulence are the strongest. Dissolved salts come via ocean currents and as the result of recycling of solids in the reef. Consequently, the most rapid growth is on the windward and currentward side of a coral island, atoll, or other land body.

The topographic profile of Fig. 5–6 shows the major elements of the windward side of a coral reef in an atoll. The prevailing winds and waves agitate the water, so that it contains a substantial concentration of dissolved carbon dioxide. The incoming ocean current brings a constant supply of dissolved salts as nutrients. Hence the reef facing the current is the most actively growing part. The *algal ridge* lies slightly awash at low tide. Most of it is algal and coralline material of compact, rugged architecture suited to the high-energy environment in the zone of wave action.

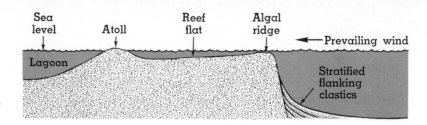

FIG. 5-6. Profile of a coral atoll, windward side.

Because the most rapid growth is in the photic zone, an overhang may and often does exist. The structure is porous and may break under its own weight, the fragments falling downward and building a clastic deposit with stratification near the angle of repose and dipping seaward from the main reef. The combination of original material, skeletal elements of other organisms living in the open spaces in the reef, and the deposition of clastic sediment composed of crushed and broken plant and animal skeletons eventually forms a structureless, massive, and homogeneous but porous and permeable reef core.

FIG. 5-7. The delicate branching or staghorn corals characteristic of the low-energy level of the reef flat. (The American Museum of Natural History, New York.)

Behind the algal ridge is the *reef flat*, an area of shallow water. Protected by the algal ridge, the water contains less carbon dioxide and the energy level is lower, so different organisms populate this part of the reef (Fig. 5–7). Some are of delicate form; others are scavengers in habit. Wave action throws clastic inorganic and organic debris from the algal ridge into the reef flat, where it makes up finer sediment and food for the animals living on and in the bottom. Processes are in the main degradational. The *island*, behind the reef flat, is composed of more-or-less stratified clastic sediment piled up by waves that have moved across the reef flat. Although the sediment is truly a clastic, all the particles are of organic origin. The algal ridge, the reef flat, and the island each range in width from a few feet to several hundreds of feet, dimensions increasing as the reef grows. Boundaries are not exact, being modified by irregularities of growth, erosion, and the tides.

The *lagoon* behind the island is a more quiet environment, since it is further shielded from wind and wave action. Most sediment is reworked material from the island. If the profile of Fig. 5–6 were to be extended to include the leeward side, the principal difference would lie in the fact that, since the atoll was developed by a unidirectional wind and current system, reef growth would be substantially less than on the windward side. Except for importation of nutrients by ocean currents, reefs can be considered to be balanced communities living in a dynamic equilibrium.

Some fossil reefs are rich reservoirs of petroleum. In such instances the original content of organic matter may have been the source material of oil, the high porosity and permeability provided for its accumulation and production, and overlying sediment trapped it in the reef.

The occurrence together and similarity of basic elements of fringing reefs, barrier reefs, and atolls have suggested to many geologists that the three types have a common origin and are steps in a continuous developmental sequence. One theory that has been proposed to explain their origin, growth, and change in geometry is based on a relative rise in sea level and depression of the ocean bottom at such a rate that the elevation change is equal to or less than the upward growth of reefs in a particular environment. The steps are as indicated in the diagrammatic sketches of Fig. 5–8. In stage I, a fringing reef more-or-less completely surrounds an island, as indicated by the shaded areas on the map and cross section. Stage II is reached after relative rise of sea level (whether by actual water-level rise or by subsidence of the crust) and growth of the reef upward and outward. The fringing reef then becomes a barrier reef. The island becomes smaller as more of it becomes submerged. A lagoon lies between the reef and the island. In stage III the island has been completely submerged and only the atoll remains, surrounding a lagoon.

The lagoons of many atolls in the Pacific Ocean have nearly flat bottoms

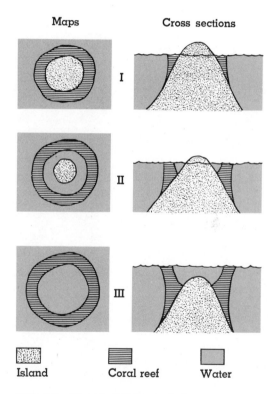

Maps   Cross sections

I

II

III

Island   Coral reef   Water

FIG. 5-8. Steps in development of a coral reef (vertical exaggerated).

of about the same depth. This is the basis for a theory linking coral reefs to continental glaciation, with the sea-level rise explained by the melting of ice sheets. But not all atolls fall within the general range of properties, and it is possible that the cause of guyots—i.e., ocean-bottom subsidence—may be related to coral reef origin.

NEGATIVE FEATURES OF THE OCEAN FLOOR

Because the ocean bottom is the site of sedimentation, we can expect that only large depressions, depressions that form more rapidly than sediment fills them, and depressions that are the result of local processes will be seen. Among these the cleft along the midocean ridges is an example of the first category and *submarine canyons* are thought to be in the last group. Submarine canyons may exist on the continental shelf or the continental slope, or may extend unbroken from one to the other (Fig. 5–9). They are to be found on all continental margins. Some are more-or-less obviously continuations of large rivers on the land surface, but some bear no relationship to the surface drainage system. Within the accuracy of present methods of subaqueous surveys they appear to range in size up to dimensions comparable with the Grand Canyon of the Colorado; at least one is 5000 feet deep at the deepest known place and 150 miles long. Materials exposed in them include recently deposited uncemented sediments, stratified sedimentary rocks, and basaltic igneous rocks.

Several geometric features of submarine canyons appear to have significance for theories as to their origin. Map patterns resemble patterns of youthful rivers in sinuosity and in the existence of few tributaries. Profiles across canyons are V-shaped, and long profiles are concave upward (steeper upslope than downslope) but are steeper than most river valleys of similar size and form. Deltaic, or fan-shaped, bodies of sediment occur at the downslope ends of some canyons. The Congo Canyon has distributaries at its lower end. A number of canyons are so accessible that they have been examined in detail at frequent intervals. Some on the California coast of North America are known to fill with sediment slowly and then empty rapidly. Canyons have actually been observed with sediment flowing down the canyon walls in elongate, streamlike bodies.

One theory of the origin of submarine canyons holds that subaerial erosion on the continental shelf and slope cut the canyons at a time when sea level was lower than it is now. Disadvantages of this hypothesis are the too-steep gradients, the lack of stream terraces, and the need for great fall of sea level without a place for the corresponding volume of water to go. A second, more likely theory is that the abrasive action of subsea currents carrying clastic sediment was responsible for canyon erosion. The observed filling and emptying of submarine canyons is substantiating evidence, as is the sediment in fan-shaped bodies below some canyons.

FIG. 5-9. Part of San
Lucas Canyon, a
submarine canyon off the
southern end of Baja
California. (U.S. Navy.)

Island arcs and the associated trenches are recently formed crustal features that have not yet been reduced to unrecognizable surface irregularities. The great depth of the trenches is related to their youth, for they have not yet been filled with sediment.

MOVEMENTS OF OCEAN WATER

CURRENTS

Dissolved and particulate matter as well as heat are transported in the oceans by currents of water. Because several kinds of motive power are available, different sets of currents are found. Some act over great areas or great depths; others are more restricted in extent.

The term *current* is applied to the movement of a large body of water of composition and/or temperature measurably different from the main body of water through which it flows with a more or less determinable velocity, direction, path, extent, and boundaries. Currents are usually linear in form but may also be broad and sheetlike. Some are restricted to horizontal motion at a constant depth, as at the surface; some follow the ocean-bottom topography to great depths, and others have a large vertical component of flow. All directions of flow have been observed. Temperatures and salinities of currents range through almost the full span known in ocean waters.

Most currents appear to be in dynamic equilibrium and presumably have existed for very lengthy periods. Some are seasonal in duration and others (upwellings, for example) occur in patterns as yet unclear to us. For surface currents our knowledge is extensive, as data have been collected since man first began sailing the seas, but the subsurface currents are not so well understood. Instrumentation and the location and control of devices are critical problems. A floating bottle stays at the surface and is comparatively easily recovered, even by someone entirely ignorant of oceanography, but it is quite a different matter to place an object in a current that flows in the vertical dimension, to keep it in the current at no matter what depth and location, and to accommodate all essentials of the device in a compact, economical, and long-lived assembly. Even the problem of determining the depth of an instrument becomes difficult without following each instrument with a ship capable of ranging the object. Automatic devices are not reliable because of the complexities of the independent and dual controls of salinity and temperature in the depth–pressure–specific-gravity relationship.

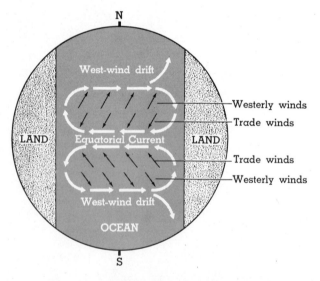

FIG. 5-10. Diagrammatic representation of the wind-derived ocean currents. Black arrows represent part of the planetary wind system and white arrows the ocean surface currents.

Currents may originate in several ways. The primary causes of the major water movements are (1) the force of gravity, acting on water masses of different densities, and (2) wind friction, acting on the ocean surface.

Density differences are caused by heating, precipitation, melting of ice, and inflow of river water, all of which reduce the density of the sea water; and by cooling, evaporation, and formation of ice, which increase the density. In the general worldwide picture, water of lesser density is formed in equatorial regions, primarily because of warming and in spite of greater evaporation; and water of greatest density is formed at high latitudes, primarily because of cooling.

If there were no forces other than gravity acting, the density differences would result in a fairly simple circulation of oceanic water from the equatorial region to high latitudes at the surface and return currents toward the equator at depth. Wind friction, however, alters the direction of flow of surface currents, and sets up *wind-friction currents*. Figure 5–10 shows with black arrows part of the planetary wind system and with white arrows a part of the surface current system. The continuously blowing winds set up permanent currents in the oceans.

The directions of ocean currents are further altered by the Coriolis effect, due to the rotation of the earth (Chapter 3), and the deflecting effect of the land masses and bottom topography. The simple diagram of Fig. 5–10 does not fully reflect reality because bottom topography, continent outline and position, and other currents exert some control over surface currents.

Most surface-current velocities are less than 2 miles per hour. The Gulf Stream (Fig. 5–11), which flows along the eastern seaboard of the United States, crosses the North Atlantic Ocean, and warms northwestern Europe, is one of the most interesting surface currents. For much of its existence it is 50 to 150 miles wide and 1500 to 5000 feet thick; it flows at an average velocity of 3 miles per hour and up to 6 miles per hour in the threads of maximum velocity. About 70 million tons of water per second flow past a given point. The water is much warmer than the landward Atlantic Ocean, and the dividing line between the current and the adjacent water is in some places very sharp. One observer reports measuring a 20-degree-Fahrenheit difference between water at the bow and the stern of a ship crossing the boundary. Its comparatively great velocity and volume result from northward diversion of much the greater part of the Atlantic Equatorial Current by the unsymmetrical position and form of South America on and near the equator. The easternmost point lies south of the equator and shunts most of the Equatorial Current northward, the smaller part being directed southward into the South Atlantic Ocean. In the North Atlantic, the Gulf Stream blends into the West-wind drift, which divides, part flowing northeastward past the Scandinavian penin-

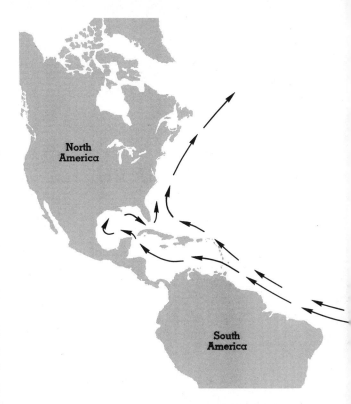

FIG. 5-11. Course of the Gulf Stream and related currents, simplified. Velocity is indicated by length of arrows.

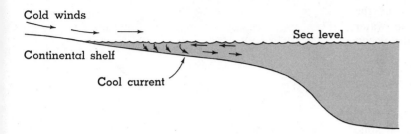

FIG. 5-12. Cross section showing convectional currents of cool water.

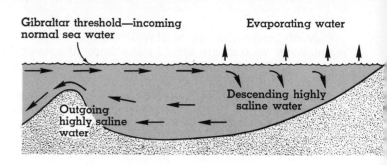

FIG. 5-13. Diagrammatic cross section showing salinity currents in the Mediterranean Sea.

sula and part flowing southward toward the equator, where it continues westward, completing a great *gyre*, or subcircular continuous path. Other currents form similar gyres in other oceans.

Density currents, as noted earlier, are formed by water in bodies with different density, or specific gravity, from that of the surrounding water, and one reason for an increase in specific gravity is cooling and resultant contraction. This process occurs along continent shores in mid-latitudes during the winter. Winds blowing off the continents cool the surface water, which shrinks, settles to the bottom, and flows seaward beneath the surface water of higher temperature and lower density (Fig. 5-12). Permanent currents of this origin are established in the polar regions and return to the equator at depth water that flowed poleward at the surface. A deep, cold current from the Antarctic flows equatorward on the west side of the South Atlantic below the depth of 12,000 feet, crosses to the east side of the Mid-Atlantic Ridge through a saddle near the equator, and continues to flow northward on the east side of the North Atlantic.

Another type of density current is formed by an increase in salinity that raises the specific gravity of the water. The Mediterranean Sea is an excellent example of such a system. It is near the subtropical high, where warm, dry, descending air establishes an environment in which evaporation dominates. Evaporation is so rapid that it is in part responsible for the fact that the surface of the Mediterranean Sea slopes downward to the east. Total evaporation exceeds stream inflow, so water must enter from the Atlantic Ocean to replenish that lost to the atmosphere. As evaporation continues, water in the Mediterranean Sea becomes more saline (about 10 per cent saltier than normal sea water) and more dense, and sinks to the bottom. Current flow is shown in Fig. 5-13. Normal marine water flows in at the surface from the Atlantic Ocean at Gibraltar, where there is a threshold on the bottom. The Mediterranean basin is filled with high specific gravity salt water, which spills over the lip of the basin below the incoming water of normal salinity. The more dense water flows down the sloping bottom of the Atlantic Ocean until it reaches about 6000 feet, the depth of the cold current of Antarctic origin, mentioned on p. 122. At that depth the two have about the same density, one because of temperature and the other because of salinity, and each loses its identity as they mix.

TIDES

Tides may be defined as daily cyclic changes in the elevation of the surface of the ocean at a given point. The gravitational attraction exerted by the moon and the sun is the cause of such changes. The layer of water on the earth's surface is not completely restrained; that is, it is to some degree free to move. The gravitational pull of the moon on the oceans causes a shift of water toward the point on the earth's surface nearest the moon (Fig. 5–14). A second cause of tides results from the revolution of the earth and the moon around a common center (Fig. 5–15). The moon is large enough and close enough to the earth for the combination to be called, with some justification, a double planet. As the pair moves through space in orbit around the sun, they tumble over and over as if connected by a cable or a rod. The center of gravity of each traces an undulating path around the sun, but the center of rotation of the pair follows a much smoother curve. Point A on the earth is farther from the center of rotation than point B and centrifugal force is stronger there, which causes water to shift toward the point on the earth which is opposite to the moon. As a result there are two broad crests on the ocean, one on the side of the earth toward the moon and one on the opposite side. These move around the earth, following the apparent motion of the moon. The combination of moon revolution around the earth and earth rotation on its axis establishes the tidal cycle of 24 hours and 50 minutes.

The sun, although many times larger than the moon, is so far away that its effects are much less. But when the earth, sun, and moon approach positions in a straight line (Fig. 5–16) the gravitational effects of moon and sun are added together and as a result high tides are higher and low tides are lower than the average. These are called *spring tides*. In contrast, when the sun and the moon are in positions about 90° from each other, the effect is to reduce tides (Fig. 5–17), and these are the *neap tides*.

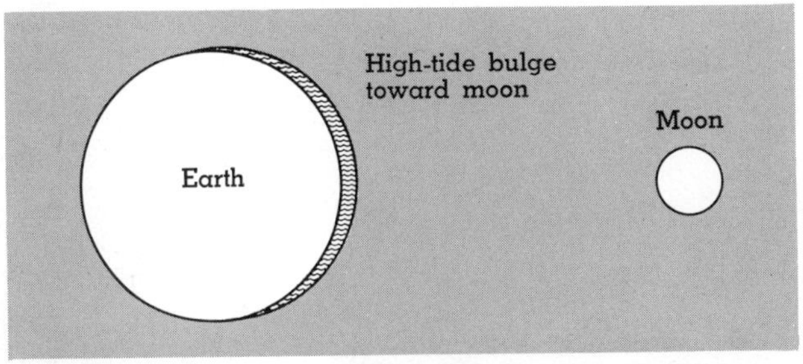

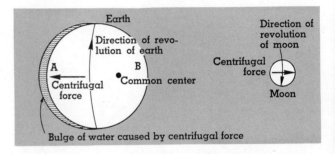

FIG. 5-14. Formation of tidal bulge by the gravitational attraction of the moon (not to scale).

FIG. 5-15. Rotation of the earth-moon system (not to scale).

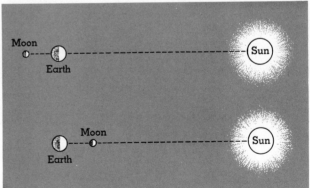

FIG. 5-16. Moon and sun reinforcing to increase tide amplitude (not to scale).

FIG. 5-17. Moon and sun in opposition, decreasing tide amplitude.

FIG. 5-18. A point on the New Brunswick shore at low tide. The water line at high tide is near the vegetated surface of the rock outcropping, some scores of feet above low-tide level. (Canadian Consulate General, New York.)

Tidal ranges differ in amount over the earth. At sea elevation differences between low and high tide are much less than in many places along the shore. As the tidal bulge moves around the earth water is actually moved laterally and tides are likely to be of different amplitude on opposite sides of a continent. The combination of the configuration of the shoreline and

124

bottom topography has the greatest effect on the range between low and high tide. A comparatively small tide far from shore, if funneled up a bay of proper form, can reach an amplitude of 50 feet, as in some bays along the New England and adjacent Canadian coast (Fig. 5–18). In similar circumstances the tide may form a wave, called a *bore*, which progresses upstream in a coastal river. These may be several feet high and move up the river for several miles at velocities as high as 12 miles per hour, with great effects on navigation in the stream.

Tidal currents are established by the ebb and flow of tides across the continental shelf, as shown in Fig. 5–19. From high tide to low tide, a volume of water is removed; this can occur in the cross section only by flow across the continental shelf and out to sea. Between low tide and high tide the reverse flow occurs. The entire volume of water flows in and out across the edge of the continental shelf each cycle; if the shelf is broad and the tidal range large, substantial currents may flow across. Currents up to 1½ miles per hour have been measured—a velocity sufficient to transport fine sand.

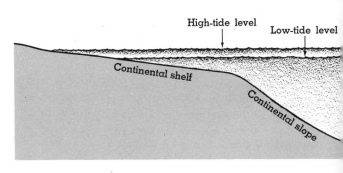

FIG. 5-19. Origin of tidal currents (vertical much exaggerated).

TURBIDITY CURRENTS

Turbidity currents are a type of density current that results from the incorporation of clastic sediment in a moving body of water. They appear to originate by disturbance of uncemented sediment on a sloping bottom. Earthquakes or storms can trigger the motion. The existence of one famous example on the Grand Banks in the North Atlantic was established by circumstantial evidence: Several transatlantic cables were broken within a few hours; as the locations and time sequence of the breaks were known, the velocity and direction of movement could be calculated. The turbidity current reached a maximum average velocity of almost 55 miles per hour. A segment of one cable was transported several miles by the current.

It is thought that where turbidity currents are significant geologic agents, sediment on the ocean bottom is stratified in units possessing *graded bedding*—the property in which sediment diameter decreases upward within a bed.

EFFECTS OF CURRENTS

The results of the action of currents are primarily associated with transportation. Clastic particles are transported in density currents and deposited in other places on the continental slope, continental rise, or the deeper basins. Tidal currents remove finer particles from the continental shelves, taking some seaward down the continental slope and depositing them out of reach of any but turbidity currents, and leaving residual deposits with higher percentages of coarser sediments. Microplankton in large numbers are transported great distances, sometimes to areas where fish live; they are then available for food as the currents mix. Commercial fisheries often exist at such sites. Icebergs are moved equatorward by cold surface currents and are of particular significance as hazards to navigation, especially as they are associated with fogs caused by the effects of the same currents on the weather. Dissolved salts and gases in currents are nutrients for a wide variety of life forms, and where currents of different compositions mix, each contributing its special elements and compounds, life may be so flourishing as to create rich fishing grounds.

Ocean currents are also of great importance in determining the climates of the world, as we shall see in the next chapter. The Gulf Stream is especially conspicuous in this regard. The heat stored in it as the water flows westward along the equator is released to warm northwestern Europe to a degree otherwise unexpected for the latitude. Other currents have similar, if less marked, effects on weather and climate through their high or low temperatures.

SUGGESTED REFERENCES

Bascom, Willard: *Waves and Beaches: The Dynamics of the Ocean,* Doubleday & Company, Inc., Garden City, N.Y., 1964. (Paperback.)

Chapin, Henry, and F. G. Walton Smith: *Ocean River,* Charles Scribner's Sons, New York, 1952. (Paperback.)

Coker, R. E.: *This Great and Wide Sea: An Introduction to Oceanography and Marine Biology,* Harper & Row, Publishers, Inc., New York, 1962. (Paperback.)

Daniel, Hawthorne, and Francis Minot: *Inexhaustible Sea,* Collier Books, New York, 1961. (Paperback.)

Defant, Albert: *Ebb and Flow: The Tides of Earth, Air, and Water,* University of Michigan Press, Ann Arbor, Mich., 1958. (Paperback.)

Long, John C.: *New Worlds of Oceanography,* Pyramid Publications, Inc., New York, 1965. (Paperback.)

Sverdrup, H. U., M. W. Johnson, and R. H. Fleming: *The Oceans, Their Physics, Chemistry, and General Biology,* Prentice-Hall, Inc., Englewood Cliffs, N.J., 1942.

Wiens, H. J.: *Atoll Environment and Ecology,* Yale University Press, New Haven, Conn., 1962.

6

We have said a great deal thus far about weather—the basic weather elements and the controls that bring about their variation in time and place—but until we discuss climates we have said nothing about the circumstances under which men actually live. For climate is the sum total of weather elements that characterize a given region of the earth's surface. As man extends the boundaries of his realm of exploration—and perhaps ultimately of colonization—far beyond the planet, we sometimes forget the extent to which he must still reckon with the natural conditions that govern the different regions of the earth.

The various climates of the earth differ greatly, and their effects upon other aspects of the physical environments in which they prevail is very great. Climate influences, to a large degree,

> the nature and amount of surface and subsurface water,
> the behavior of moving ice and winds,
> the manner and speed with which rocks decompose or disintegrate,
> the type of natural vegetation that prevails,
> the range of crops that can be produced, and
> the range of insects, bacteria, and viruses present.

And the climates of the past account for the presence of deposits of coal, salt, gypsum, limestone, etc.

No other aspect of the physical environment has such a great impact, direct or indirect, on the character of a region.

CLASSIFICATION OF CLIMATES

The climates of the world have in the past been grouped in a few very general categories: on the basis of temperature—tropical, temperate, and polar; on the basis of precipitation—humid, subhumid, semiarid, and arid; and on the basis of vegetation—tropical forest, taiga, prairie, steppe, and so on. These categories are so general that they have limited value, either practical or scientific. Five decades ago an Austrian scientist, Wladimir Köppen, devised a detailed system of classification designed to meet the need for more precise designations. As revised over the years, it has become widely accepted. Monthly rainfall and temperature data are used in combination to differentiate climate groups and subgroups. There are five major categories:

> *A* climates: tropical temperatures and plentiful rainfall
> *B* climates: arid and semiarid
> *C* climates: humid; mild, or mesothermal, winters
> *D* climates: humid; severe winters and mild summers
> *E* climates: polar, no warm season

Tropical humid climates (A)

The *A* climates have one outstanding characteristic: they lack a cold season. The average temperature for every month is over 64.4° F. (Table 6–1). If one excludes the vast arid and semiarid zones of the world, the tropical humid climates occupy more land area than any other category. These climates spread poleward from the equator at low elevations to as much as latitude 25° along the windward sides of the continents. However, there may be considerable variation within these regions, depending upon local wind systems, continental distribution, and topography.

The *A* climates are divided into 2 subgroups based on precipitation: (1) The tropical rainforest climates (*Af, Am*) have abundant rainfall throughout the year, thus no real dry season. (2) The savanna climates (*Aw, As*) generally have less total rainfall, and precipitation is concentrated in one season; hence there is a prolonged dry season.

TROPICAL RAINFOREST CLIMATE (Af, Am)

This climate is distinguished from all others in that it is consistently warm and consistently wet. The *Af* climate has at least 2.4 inches of rainfall for every month. Although the *Am* climate may have one or more months with less, the total amount for the rest of the year is sufficient to offset the brief dry period. These regions are on or near the equator, and they experience high angles of the sun throughout the year and consequently little variation in the length of day. Thus average monthly temperatures are consistently high; typically, 75 to 80° F. The annual ranges of temperature are very low, from 0° to less than 5° (Fig. 6–1). Water vapor content of the air is high, causing moderate daily ranges of temperature—15 to 25°. But note that the daily ranges are high as compared with the annual ranges. This fact is well expressed in the saying, "Night is the winter of the tropics." The high relative humidity, intense sunlight, and small amount of wind cause very high sensible temperatures. Although the actual temperatures are not excessive (85–90° F. during the day and 65–75° F. at night), the weather is experienced as sultry and oppressive.

Rainfall is heavy—from 70 to 100 inches or more per year. It is convectional in origin and is usually in the form of short showers, with an abundance of cumulus and cumulonimbus clouds. Thunderstorms occur more frequently in these regions than in any other, some localities consistently experiencing 200 or more thunderstorm days annually. The winds in the *Af* climates are variable and weak owing to conditions in the equatorial low, or doldrums belt. The *Am* climates in southeastern Asia experience a reversal of winds during the year as a result of the monsoonal system of air circulation.

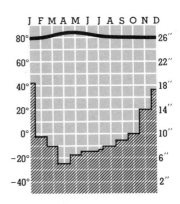

FIG. 6-1. *Af* climate—Sandakan, Maylasia. Annual temperature, 81°; annual precipitation, 119.7". The cold month is above 64.4° and each month of the year receives at least 2.4" of rainfall.

TABLE 6-1   KÖPPEN SYSTEM OF CLIMATE CLASSIFICATION

A: Tropical climates—average temperature of coldest month is above 64.4° F.

   f   Every month has 2.4″ of precipitation or more.
   m Short dry season. Resembles *Af* in amount of precipitation but *Aw* in seasonal distribution. The limit between types *Am* and *Aw* is $a = 3.94 - r/25$, where *r* is annual rainfall and *a* is the rainfall of the driest month, both in inches (see Table 6-2).
   w Well-defined dry season in winter.
   w′ Rainfall maximum in autumn.
   w″ Two distinct rainfall maxima separated by two dry seasons.
   s   Well-defined dry season in summer (very rarely found).
   i   Isothermal; yearly range of monthly average temperatures is less than 9° F.
   g   Warmest month before summer solstice.
   t′ Warmest month delayed until fall.

B: Arid and semiarid climates—evaporation exceeds precipitation. The boundary between the dry and humid climates is defined as follows (see also Table 6-3):

   $r = 0.44T - 8.5$ when rainfall is well distributed throughout the year,
   $r = 0.44T - 3$ when 70% or more of the rain falls in the summer,
   $r = 0.44T - 14$ when 70% or more of the rain falls in the winter,

   where *T* is temperature.

   S Steppe ⎫
   W Desert ⎬ The *BS–BW* boundary is *r* in the above equations, divided by 2.
   h   Annual temperature is 64.4° F. or above.
   k   Annual temperature is less than 64.4° F.
   s   70% or more of total rainfall in the winter 6 months.
   w 70% or more of total rainfall in the summer 6 months.

C: Mesothermal climates—average temperature of coldest month is under 64.4° F. and above 26.6° F.

   f   Precipitation does not meet conditions of either *w* or *s*.
   s   Wettest winter month is three times as wet as driest summer month, and driest summer month has 1.6″ of rain or less.
   w Wettest summer month is ten times as wet as driest winter month.
   a   Warmest month is over 71.6° F., and at least four months are over 50° F.
   b   No month above 71.6° F., but at least four months are above 50° F.
   c   One to three months above 50° F.
   t′ Warm month delayed until fall.
   x   Precipitation maximum in spring or early summer.
   i, g Same as in *A* climates.

D: Microthermal climates—average temperature of coldest month is under 26.6° F. Subheadings under this classification are the same as in *C* climates, except for the following tertiary letters:

   c   One to three months above 50° F., with coldest month above −36.4° F.
   d   Average temperature of coldest month is −36.4° F. or under.

E: Polar climates—average temperature of warmest month is below 50° F.

   T Average temperature of warmest month above 32° F. and under 50° F.
   F Average temperature of warmest month is under 32° F.

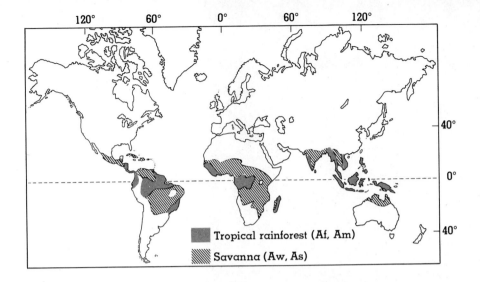

Tropical rainforest (Af, Am)

Savanna (Aw, As)

FIG. 6-2. Generalized map showing distribution of the tropical humid climates: tropical rainforest (*Af*, *Am*) and savanna (*Aw*, *As*).

The tropical rainforest climates are found at low latitudes and at elevations under 3000 feet. (At higher altitudes temperatures are lower than 64.4° F. for the colder months, and hence do not meet the criteria for the *A* climates.) They cover about 8 per cent of the land area of the earth and are occupied by about 5 per cent of the world's population. They are found 5–10° on either side of the equator but extend poleward to latitude 25° on windward coasts (Fig. 6–2). In Latin America, they are in the Amazon basin (the largest contiguous area of *Af, Am* climates in the world), the Guyana coast, the Bahia coast, the Caribbean coast of Central America, windward coasts of certain islands in the West Indies, and the Pacific coast of Colombia. In Africa, the largest extent of this climate is in the Congo basin. It is also found along the Guinea coast and on the eastern side of the island of Madagascar. The East Indies are dominated by this climate, and it extends poleward along the Malayan and Burmese coasts to the Ganges delta. It is also found along the Malabar coast of India and in the southern third of Ceylon.

The high temperatures and lack of a prolonged dry season produce rapid plant growth during the entire year. Rainfall is sufficient to support a forest type of natural vegetation. The *Af* climates are associated with dense, high evergreen tropical rainforests (Fig. 6–3), and the *Am* climates with a lower and less dense forest cover. Because heavy rainfall causes excessive leaching (removal of soluble constituents by water percolating downward), the topsoil is usually deficient in key mineral plant foods and organic materials. The unproductive residual soil and the widespread distribution of biotic pests present formidable obstacles to agriculture.

FIG. 6-3. Selva, or tropical rainforest, Canal Zone. Note the great number and variety of species of plants typical of the selva. This type of forest is evergreen and is adapted to high-moisture conditions. (The American Museum of Natural History, New York.)

EARTH SCIENCE

The savanna climate has tropical temperatures throughout the year and a well-defined wet season followed by a prolonged dry season. The amount of rainfall in the dry period determines whether a locality is classified as *Am* or as a savanna (*Aw* or *As*) climate (Table 6–2). The dry season occurs during the period of low sun in the *Aw* climates (*w* denotes winter dry) and in the high-sun period in the *As* (*s* for summer dry) climates. The *As* type is far less widespread and is found primarily in southeastern Asia, in areas located in a rain shadow of the summer monsoon and where the winter monsoon ascends orographically. Temperature conditions are similar to those of the tropical rainforest climates except that the annual range is greater—it is generally 5–20°—reflecting the location farther from the equator. Typically, the coldest month averages 65–75° F. and the warmest month 80–90° F.

Total annual rainfall is usually about 40–70 inches, concentrated in one season, but in atypical areas it may be less than 35 inches or over 400. The seasonal distribution of rain may be the result of either of two conditions. In certain areas, the latitudinal migration of the wind and pressure belts causes the weather to be dominated by the doldrums, with heavy rain, during one season and the trade winds, with associated drier conditions, during the other season. In areas influenced by the monsoonal system of winds, a moist wind is experienced in one season and a dry wind from the opposite direction during the other. Regardless of the cause, however, the most important distinguishing characteristic of the savanna climates is the existence of alternating wet and dry seasons (Fig. 6–4).

Savanna climates cover approximately 14 per cent of the earth's land areas and are inhabited by about 15 per cent of the world's peoples. They are found from latitudes 5° to 20° and may extend poleward to latitude 30° along the windward sides of continents (Fig. 6–2). In Latin America, they occupy a vast area in Brazil known as the *campos*; and in the Northern Hemisphere, an area of the Orinoco River valley called the *llanos*. Much of the Pacific coast of Central America and the West Indies falls within this climatic category. In Africa, the savanna climate is found in the Sudan region north of the equator and the Veld region south of the equator, as well as in central and western Madagascar. Most of peninsular India and southeastern Asia and northern Austrialia have this type of climate.

It should be noted that the savanna climates lie poleward of the tropical rainforest climates. The *Af, Am* climates straddle the equator and fall within the doldrums for most, if not all, of the year, consequently receiving abundant rainfall, whereas the savanna climates are usually on the margins of the doldrums and the trade wind belts. Since these belts migrate latitudinally, the savanna climates are under the influence of the doldrums part of the year and the trade winds the other. The drought season is

TABLE 6–2 DETERMINATION OF THE *Am* AND *Aw* OR *As* BOUNDARY[a]

TOTAL ANNUAL RAINFALL (in inches)	RAINFALL OF DRIEST MONTH
40	2.34
42	2.26
44	2.18
46	2.10
48	2.02
50	1.94
52	1.86
54	1.78
56	1.70
58	1.63
60	1.55
62	1.47
64	1.38
66	1.30
68	1.22
70	1.13
72	1.06
74	0.98
76	0.90
78	0.81
80	0.74
82	0.66
84	0.58
86	0.50
88	0.42
90	0.34
92	0.26
94	0.18
96	0.09
98	0.02

[a] Table modified from Kendall, **H. M.**, et al., *Introduction to Geography*, 4th ed., 1967, by permission of Harcourt, Brace & World, Inc.

Note: If rainfall of the driest month is more than is shown in second column the symbol *m* is used; if less, the symbol *w* or *s* is used.

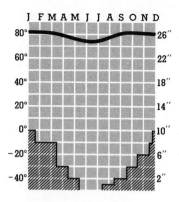

FIG. 6-4. *Aw* climate—Cuyaba, Brazil. Annual temperature, 79°; annual precipitation, 53.6". The cold month is above 64.4° and there is a distinct dry season during the period of low sun.

generally long enough to prevent the growth of dense forests, and the typical natural vegetation is tall tropical grasses and small, widely spaced trees. Certain areas are densely populated—generally those with productive soils—and others very sparsely populated.

ARID AND SEMIARID CLIMATES (*B*)

The dry climates cover a greater portion of the world's land area than any of the other categories. They are found from within a few degrees of the equator to almost latitude 55° in both hemispheres. The *B* climates have one characteristic in common—they all have less precipitation than evaporation; but because of the great latitudinal extent within which they occur, other climatic characteristics may be widely divergent. The interaction of the wind and pressure belts, the mountain barriers and land masses, and the ocean currents cause marked climatic variations.

The *B* climates are divided into the arid *deserts* (*BW*) and the semiarid *steppes* (*BS*). The basis for this division is the total annual amount and seasonal distribution of rainfall, and the temperature as a major regulator of evaporation. The amount of rainfall that separates *BS* and humid climates is shown in Table 6–3. It is important to bear in mind that evaporation is taken into consideration in addition to the amount of rainfall. Thus a tropical area having 30 inches of rain annually may be classified as a semiarid steppe; whereas in midlatitudes, where lower temperatures prevail, and hence less evaporation, the same amount of rain would be more than sufficient to classify the area as a humid-climate region. The *BS* and *BW* climates are differentiated by exactly one-half the amount of rainfall that separates the *BS* and humid climates.

In addition to the division by rainfall, *B* climates are distinguished by whether they are tropical (no cold season) or midlatitude (cold winter). The average annual temperature of 64.4° F. serves as the dividing point in this respect. Thus there are four subgroups of *B* climates:

BWh: Tropical desert—arid (evaporation greatly exceeds precipitation); no cold season
BWk: Midlatitude desert—arid; cold season
BSh: Tropical steppe—semiarid (evaporation slightly exceeds precipitation); no cold season
BSk: Midlatitude steppe—semiarid; cold season

TROPICAL DESERTS AND STEPPES *(BWh, BSh)*

The tropical, or low-latitude, deserts and steppes experience annual ranges of temperature that are large for the latitude. Average temperatures for the warmest month are usually 85–90° F. or more, and the cold-month

TABLE 6-3   BOUNDARY BETWEEN *BS* AND HUMID CLIMATES[a]

AVERAGE ANNUAL TEMPERATURE	RAINFALL 70% OR MORE IN WINTER 6 MONTHS	RAINFALL EVENLY DISTRIBUTED	RAINFALL 70% OR MORE IN SUMMER 6 MONTHS
32	.08	5.58	11.08
34	.96	6.46	11.96
36	1.84	7.34	12.84
38	2.72	8.22	13.72
40	3.60	9.10	14.60
42	4.48	9.98	15.48
44	5.36	10.86	16.36
46	6.24	11.74	17.24
48	7.12	12.62	18.12
50	8.00	13.50	19.00
52	8.88	14.38	19.88
54	9.76	15.26	20.76
56	10.64	16.14	21.64
58	11.52	17.02	22.52
60	12.40	17.90	23.40
62	13.28	18.78	24.28
64	14.16	19.66	25.16
66	15.04	20.54	26.04
68	15.92	21.42	26.92
70	16.80	22.30	27.80
72	17.68	23.18	28.68
74	18.56	24.06	29.56
76	19.44	24.94	30.44
78	20.32	25.82	31.32
80	21.20	26.70	32.20
82	22.08	27.50	33.08
84	22.96	28.46	33.96

[a] Table modified from Kendall, H. M., et al., *Introduction to Geography*, 4th ed., 1967, by permission of Harcourt, Brace & World, Inc.
Note 1: All rainfall figures given in inches.
Note 2: Boundary between *BS* and *BW* climates is *exactly one-half* the amount of rainfall that differentiates the *BS* and humid climates as shown above.

temperatures are 50–60° F. In all cases, however, the average annual temperature is above 64.4° F. The annual ranges of 25–40° are the highest for any tropical climate. Daily ranges are extremely high, reflecting the clear skies and dry air. Daytime temperatures of over 100° F. are common, and the world's highest official temperature readings of over 136° F. have been recorded in low-latitude deserts. Nighttime temperatures of 50–70° F. are not unusual, and light frosts occur occasionally.

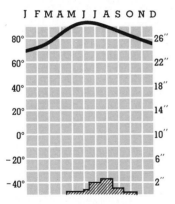

FIG. 6-5. *BWhw* climate—Khartoum, Sudan. Annual temperature, 84°; annual precipitation, 5.4″. The annual temperature is above 64.4°; evaporation greatly exceeds precipitation; and more than 70 per cent of the precipitation is concentrated in the summer months.

Tropical deserts and steppes extend to the oceans along the west coasts in many areas. These deserts and steppes experience temperatures 10–20° lower than interior locations. The tropical deserts are the driest regions of the earth—rainfall for the year normally varies from less than 1 inch to about 10 inches (Fig. 6–5). The low-latitude steppes are less dry, with total annual precipitation of 10–25 inches as a rule. The poleward margins of the dry-climate regions usually do not receive any rain in the summer, when they are under the influence of the subtropical high, but receive some cyclonic rainfall associated with the belt of westerlies in the winter. Steppes closest to the equator receive most of their precipitation in the summer season, when the doldrum belts have migrated to their furthest point poleward. Tropical dry climates that receive 70 per cent or more of their total rainfall during the winter are classified as *BWhs* or *BShs* (*s* denoting summer dry), and those receiving 70 per cent or more of the total rainfall in the summer as *BWhw* or *BShw*. Tropical dry climates generally experience 75–90 per cent of possible sunshine, and minimum cloudiness. However, this is not true in areas along the west coasts where cool ocean currents are adjacent to land. Although rainfall is light here, fogs and low stratus clouds are abundant.

The tropical dry climates result from one or more of the following factors: (1) the influence of the subtropical high, in which currents

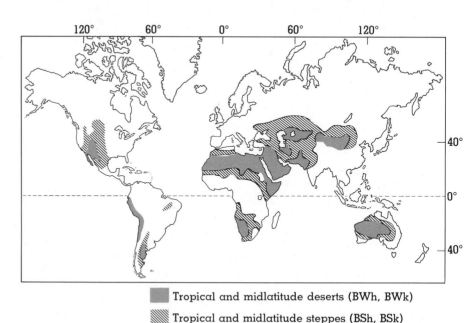

FIG. 6-6. Generalized map showing distribution of the arid and semiarid climates: tropical desert (*BWh*), tropical steppe (*BSh*), midlatitude desert (*BWk*), and midlatitude steppe (*BSk*).

▨ Tropical and midlatitude deserts (BWh, BWk)

▧ Tropical and midlatitude steppes (BSh, BSk)

descend and winds diverge; (2) the trade winds, which are dry unless forced to ascend; (3) a location in the interior of a land mass, a great distance from the source of moisture; (4) a location on the leeward side of highlands; and (5) a coastal location near cool ocean currents, which reduce rainfall (although fogs are common).

The low-latitude dry climates are located poleward of the savanna climates; they are generally at latitudes 15° to 30° and may extend latitudinally along the leeward coasts an additional 5–10° (Fig. 6–6). They attain their greatest areal extent in Africa. The Sahara region of North Africa and beyond the Arabian plateau into India is the largest contiguous region of tropical desert in the world (Fig. 6–7). The Kalaharie Desert,

FIG. 6-7. Drifting sand, typically associated with low desert (*BWh*) climates. Saudi Arabia. Sparsity of vegetation is the result of both the drifting sand and low annual rainfall. (Arabian American Oil Company, New York.)

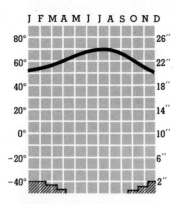

FIG. 6-8. *BSks* climate—San Diego, California. Annual temperature, 61°; annual precipitation, 9.6". The annual temperature is below 64.4°; evaporation slightly exceeds precipitation; and more than 70 per cent of the total precipitation is concentrated in the winter months.

south of the equator, is not as extensive, owing to the narrow land mass and the absence of an adjacent continent. At the same latitudes, low-latitude dry climates are found in northern Mexico and southwestern United States. In South America, the Atacama Desert is restricted to a narrow coastal strip west of the Andes Mountains. Despite its narrowness, this desert extends over a length of 2000 miles, in all of which the annual rainfall is rarely over 1 inch. Most of central and western Australia lies within the zone of low-latitude deserts and steppes.

MIDLATITUDE DESERTS AND STEPPES *(BWk, BSk)*

The midlatitude arid and semiarid climates are within the humid westerlies and are the result of locations in the lee of highlands or deep in the interior of large land masses. The most arid of the midlatitude deserts are located in basins surrounded by highlands. Because of their extensive latitudinal range, temperatures vary greatly. However, the warmer months usually average 70–80° F., and the colder months 10–40° F. (Fig. 6–8). In most of these areas there is a conspicuous cold season, which lowers the annual temperature to 64.4° F. or less. The midlatitude deserts are not as rainless as the tropical deserts, for they do receive a little precipitation each year. Interior locations tend to receive their rainfall in the summer season, and western and equatorward locations chiefly in the winter half of the year. The midlatitude dry climates extend from about latitude 20° to as far as latitude 55° in the interior of continents. The largest region of *BWk* and *BSk* climates is found in Asia, extending from the Black Sea to Mongolia (Fig. 6–6). In North America, they occupy most of the land area east of the Sierra-Cascade Mountains extending into the Great Plains (Fig. 6–9). In South America, the Patagonian Desert, east of the Andes, is the only extensive area of midlatitude desert. Because of the narrowness of the land mass, the Patagonian Desert does not experience the great annual range of temperature that exists in central Asia and central North America.

The dry climates occupy about 27 per cent of the world's land area and generally coincide with regions of sparse population. The total area of the climates is over three times the total cultivated area of the world. Only about 2 per cent of the dry-climate land area is used intensively, generally regions where irrigation is possible. In view of the small amount of precipitation and its great variability from one year to the next, it is unlikely that intensive use of large areas within the dry climates will occur in the foreseeable future. Where irrigation is feasible, high yields of a great variety of agricultural crops can be attained because of the high per cent of sun and the generally good soil, rich in mineral plant foods. Desert areas where irrigation is not carried on are sparsely populated and unproductive. The subhumid steppe climates are associated with sparse steppe

FIG. 6-9. Saguaro cactus, Arizona. Typical xerophytic vegetation adapted to drought conditions. This variety of cactus is only one of many associated with desert (*BW*) climates. (U.S. Forest Service.)

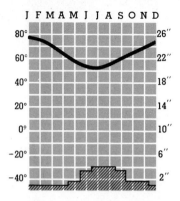

J F M A M J J A S O N D

FIG. 6-10. *Csa* climate—Adelaide, Australia. Annual temperature, 63°; annual precipitation, 21.2″. The cold month is between 26.6° and 64.4°, and rainfall is concentrated in the winter season.

grasses, which are often utilized for the grazing of livestock. Some steppes receive sufficient rainfall to produce low yields of drought-resistant grains and fibers. The dry polar and subarctic climates offer less promise for supporting large populations than any of the other major climatic types.

HUMID MESOTHERMAL CLIMATES (C)

The climates in the mesothermal group are temperate and humid. Since the limits that define this category are relatively broad, there may be considerable variation from one mesothermal climate to another. The wide statistical range is reflected in the areal patterns; although not as extensive as the dry and tropical humid climates, C climates are found in substantial areas of every continent. In general they are in the midlatitudes, although some mesothermal regions occur from subtropical latitudes to beyond latitudes 70° north and 50° south. In C climates the cold-month temperature is between 26.6 and 64.4° F. These climates are divided into three subgroups, based on rainfall distribution and summer temperature (Table 6–1): (1) Mediterranean climate (*Cs*), (2) humid subtropical climate (*Ca*), and (3) marine climate (*Cb, Cc*).

MEDITERRANEAN CLIMATE *(Cs)*

The *Cs* climate is subtropical. Summer temperatures generally average 75–80° F.; but in coastal locations near cool currents the average is 5–10° lower. Mediterranean climates are under the influence of the subtropical high in the summer, when they consequently have dry and clear weather. The summer temperatures and extremely high daily ranges are not unlike those of the deserts and steppes, which border the *Cs* climates along their equatorial margins. Typical temperature averages for the coldest month are 45–55° F. Coastal locations usually are somewhat warmer in the winter than inland locations. Frosts occur occasionally during the winter season, but the Mediterranean climates are well known for their mild winters, and many areas that fall within this climatic type have developed into major resort centers.

Precipitation is concentrated in the winter season. The wettest winter month receives at least three times as much precipitation as the driest summer month, and the driest month receives no more than 1.6 inches of rain (Table 6–1). The dry summers and mild, humid winters differentiate this climate from all other types (Fig. 6–10). Total annual rainfall is normally 15–30 inches, least along the equatorward margins and increasing poleward. The dry summers are due to the influence of the subtropical high, and the winter precipitation is associated with the westerly winds. Located on the boundary between these two belts, *Cs* climates feel the influence of each, as the belts migrate latitudinally during the year. Fogs are frequent along the coastal areas as a result of the cool currents offshore.

Mediterranean climates are found at latitudes 30° to 40° along the west coasts in both hemispheres and extending inland as far as topography will allow (Fig. 6–11). They occupy the coastal areas of central California and central Chile, in each case the high mountains trending north-south restricting their areal extent to a relatively narrow coastal strip. The southern tip of Africa and the southern and southwestern portions of Australia fall within this climate group. The greatest extent of *Cs* climate in the world is found along the borderlands of the Mediterranean Sea in southern Europe, northern Africa, and the Middle East.

The *Cs* climates cover about 2 per cent of the world's land area and are inhabited by about 5 per cent of the world's population. The subtropical temperatures permit the production of a great variety of frost-sensitive crops, such as citrus, sugar cane, cotton, and grapes. The dry summer is normally the dormant season, and irrigation is often necessary for the year-round growth of many annual plants. Some perennials, however, such as olives, cork, figs, and grapes, withstand the summer drought. Many commercial grains and out-of-season vegetables are grown during the mild and humid winter for market in poleward or inland regions,

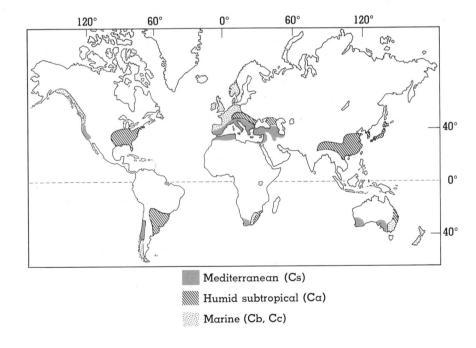

FIG. 6-11. Generalized map showing distribution of the humid mesothermal climates: Mediterranean (*Cs*), humid subtropical (*Ca*), and marine (*Cb*, *Cc*).

■ Mediterranean (Cs)

▨ Humid subtropical (Ca)

▦ Marine (Cb, Cc)

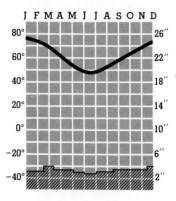

FIG. 6-12. *Caf* climate—**Buenos Aires, Argentina.** Annual temperature, 61°; annual precipitation, 37.3″. The cold month is between 26.6° and 64.4°, the warm month above 71.6°, and there is no dry season.

where the winter temperatures impose a dormant season. The small total amount of rainfall and dry summers result in a vegetative cover of stunted, widely spaced thornforest. At higher elevations, there are forests of some value and grasses for grazing.

HUMID SUBTROPICAL CLIMATE (Ca)

The humid subtropical climate differs from the Mediterranean climate in several respects: The precipitation tends to be concentrated in the warm season; the winters are usually colder; and it is located on the eastern sides of continents rather than the western coasts. Temperatures are normally 75–80° F. for the warmest month. The *a* in *Ca* denotes that the warmest month is above 71.6° F. (Table 6–1). Humidities are higher than those of the *Cs* climate, resulting in lower daily ranges of temperature and more oppressive heat. Temperatures for the coldest months range from 27 to 50° F. depending upon location (Fig. 6–12). A frost period of 3–5 months prohibits year-round agricultural production. The winter season in the *Ca* regions of North America and Asia is more severe and of longer duration than the winters in the Southern Hemisphere, where large land masses are lacking at midlatitudes.

The total precipitation averages 30–60 inches or more. As a rule rainfall decreases inland, and the driest areas are along the western margins of the climatic region. During the summer months, convectional rainfall is common and thunderstorms are frequent. In the winter season, the rainfall is more widespread and is usually associated with passing midlatitude cyclones. Snow and sleet occur occasionally but rarely remain on the ground for more than a few days.

The humid subtropical climates generally receive more rainfall in the summer season than in the winter months, but this seasonal difference varies considerably. Where the wettest summer month is at least ten times as wet as the driest winter month, the climate is further classified as *Caw*, *w* denoting winter dry. In *Caf* climates, rainfall is not markedly concentrated in either season, and the *f* denotes well-distributed precipitation (Table 6–1). *Caw* climates are located primarily in Asia, where the well-developed winter monsoon is dry and the summer monsoon brings heavy rainfall.

Humid subtropical climates are found between latitudes 25° and 40° on the east coasts of continents and extend inland as far as topography will allow (Fig. 6–11). The southeastern quadrant of the United States is representative (Fig. 6–13). Also included are most of central and southern China, southern Japan, southern Korea, and the area continuing westward along the foothills of the Himalayan Mountains into northern India. The *pampas* of South America, parts of the Balkan plateau, the northeastern

coastal strip of Australia, and the eastern coastal area of south Africa are the remaining areas of *Ca* climate.

The long frost-free season of 200 days or more; abundant rainfall during the growing season; and a short but cool winter, which retards proliferation of biotic pests, combine to make this climate the most productive in the midlatitudes. A great variety of agricultural, pastoral, and forest products

FIG. 6-13. Vegetation typical of water bodies in tropical or semitropical climatic areas. Ocala National Forest, Florida. (U.S. Forest Service.)

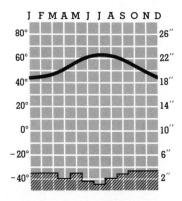

J F M A M J J A S O N D

FIG. 6-14. *Cbf* climate—Brest, France. Annual temperature, 54°; annual precipitation, 29.1″. The cold month is between 26.6° and 64.4°, the warm month is below 71.6°, but at least four months are over 50°, and there is no dry season.

are produced. In the United States South, large quantities of cotton, grains, livestock, fruits, and nuts are grown. The *pampas* of Argentina and Uruguay are known for their exports of corn, wheat, beef, mutton, wool, and flax. In Asia, rice, tea, vegetables, grains, and fruits are produced in tremendous quantities. The high productivity of the *Ca* climate regions is reflected in the fact that over one-third of the world's peoples live within their boundaries.

MARINE CLIMATE (Cb, Cc)

The marine climates are characterized by mild winters, cool summers, and well-distributed rainfall. Annual ranges of temperature are very low for the latitudes. The mild and relatively uniform temperatures and the high per cent of clouds result from the location of these climates on the west coasts. Their nearness to warm ocean currents makes it possible for the westerly winds to carry the effects of the warm water inland both in summer and in winter. Summer months are cool, with average temperatures of 60–70° F. (*b* denotes 4 months or more above 50° F. but no month reaching 71.6° F.; *c* denotes 1–3 months above 50° F.). The high water vapor content of the air results in an abundance of cloudiness, frequent fog, and small daily ranges of temperature. Winter temperatures are exceptionally mild for the latitude, averaging 27–50° F. for the cold month, and the frost-free season of 6–9 months is lengthy for the latitude.

The marine influence is the most important control of the relatively uniform day-to-day and seasonal temperatures. Since marine climates are under the influence of the westerly winds all year, rainfall is nearly uniform from season to season with often a slight maximum in the winter, and droughts are unknown. Thus marine climates are all classified as *f* in relation to rainfall distribution. These climates are probably cloudier than any others; they are characterized by widespread stratus and nimbostratus clouds and frequent advection fogs, especially in the winter season. Total annual rainfall ranges from 20 inches to over 200 inches where the orographic ascent of the westerlies occurs (Fig. 6–14).

The marine climates are located on the west coasts from latitude 40° poleward, in some cases reaching the Arctic Circle. Their greatest extent is in northwestern Europe (Fig. 6–11). Here the exceptionally warm ocean currents and the lack of a significant north-south mountain barrier allow the westerly winds to carry the marine influence far inland. Most areas in northwestern Europe are 20–45° warmer in January than is normal for the latitude. Marine climates are also found along a narrow coastal strip of southern Chile and the western coast of North America from northern California to southern Alaska. In these regions their extension inland is prevented by high north-south mountain ranges. Most of New

Zealand, Tasmania, and the tip of southeastern Australia also fall within the marine climates (Fig. 6–15).

The mild winters, long frost-free season, abundant rainfall, and lack of a dry period make possible in *Cb* and *Cc* climates the cultivation of a variety of agricultural, pastoral, and forest products. On the other hand, the cool summers and low per cent of possible sunshine are deterrents to

FIG. 6-15. Pasturelands, New Zealand. High carrying capacities of pasturelands in marine (*Cb*) climates result from mild winters, cool summers, and adequate rainfall distributed throughout the year. (New Zealand Consulate General, New York.)

the production of certain midlatitude crops; and, of course, the cold season precludes the cultivation of tropical and subtropical plants. Forage and grass crops find almost optimum conditions here, and pasturelands have very high carrying capacities. The dairy and livestock industries are well developed, especially in northwestern Europe. The original vegetation of the climatic region was primarily forest, and hilly or mountainous areas are still important sources of high-grade coniferous lumber and pulp. About 15 per cent of the world's population inhabits the marine climatic regions.

HUMID MICROTHERMAL CLIMATES (D)

The microthermal climates are characteristic of large land masses at midlatitudes. Although their areal extent is less than some other major climatic types, they occupy large areas of the Eurasian and North American continents that are densely populated and intensively cultivated. Because the limits of these climates are broad, there are several subgroups that together cover a wide latitudinal range. However, all microthermal climates possess one distinguishing characteristic—they are predominantly controlled by the large land masses. The continental extremes are most evident in any analysis of their temperature and precipitation conditions. The D climates are defined as those in which the cold month is below 26.6° F. and at least one month is above 50° F. (Table 6–1). They are found only in the Northern Hemisphere, since broad land areas are absent at midlatitudes in the Southern Hemisphere. The D climates are divided into three types: (1) humid continental warm-summer climate (Da); (2) humid continental cool-summer climate (Db); and (3) subarctic climate (Dc, Dd).

HUMID CONTINENTAL WARM–SUMMER CLIMATE (Da)

This climate is characterized by severe winters and warm summers. The cold month averages 0–26° F., the warm month must be at least 71.6° F., and the month of July usually averages 75–80° F. The Da climate of Asia is characteristically colder in the winter season than that in North America because of the larger land mass. The frost-free season of 140–200 days allows some variety in agricultural practices, but subtropical plants cannot mature in such a short period. Production of corn is widespread, particularly in the United States; hence this is often referred to as the Corn Belt climate. It is the only climate thus far considered in which a snow cover remains on the ground for an appreciable length of time. As discussed in Chapter 2, a snow cover of long duration reduces the temperature of the air considerably. Snow-covered surfaces reflect rather than absorb solar energy, so that little can be utilized in warming the ground. Also, snow is a poor conductor and it therefore prevents the heat of the ground from warming the lower layers of the air. On the other hand, a snow cover keeps the ground warm and thus prevents deep freezing in the winter.

The total precipitation in *Da* climates ranges from 20 to 40 inches (Fig. 6–16). Summer rainfall is predominantly convectional and winter precipitation primarily cyclonic. In Asia the monsoonal circulation of winds results in great seasonal extremes in temperature and precipitation. Most Asian locations are therefore classified as *Daw* and most localities in Europe and North America *Daf*, reflecting less seasonal variation in precipitation.

The *Da* climates are located in the Northern Hemisphere between latitudes 35° and 45°. They are found in the interior and on the leeward coasts of Asia, Europe, and North America (Fig. 6–17). Central and eastern United States and portions of the Balkans in Europe are representative *Daf* climates. Central China, northern Korea, and northern Japan fall within the general *Da* boundaries. Japan and coastal Asia are *Daf*, and interior Asia is *Daw*.

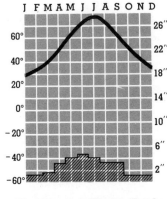

FIG. 6-16. *Daf* climate—Omaha, Nebraska. Annual temperature, 51°; range, 55°; annual precipitation, 29.0". The cold month is below 26.6°; the warm month above 71.6°; and there is no distinct dry season.

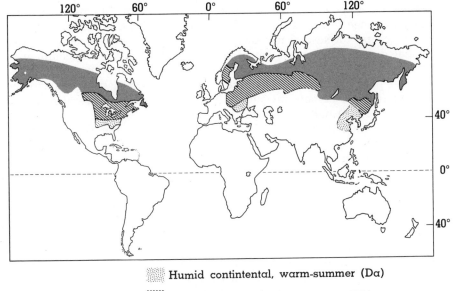

Humid contintental, warm-summer (Da)

Humid continental, cool-summer (Db)

Subarctic (Dc, Dd)

FIG. 6-17. Generalized map showing distribution of the humid microthermal climates: humid continental, warm-summer (*Da*), humid continental, cool-summer (*Db*), and subarctic (*Dc, Dd*).

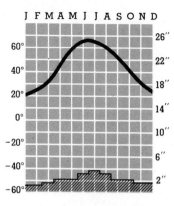

FIG. 6-18. *Dbf* climate—**Kiev, U.S.S.R.** Annual temperature, 44°; range, 46°; annual precipitation, 21.1″. The cold month is below 26.6°; at least four months are above 50°; and there is no distinct dry season.

The shorter growing season and modest amount of rainfall restrict the number of crops that can be produced in *Da* climates. Cotton, rice, tea, and citrus so characteristic of the Mediterranean and humid subtropical climates are not produced commercially. Droughts occasionally occur along the western margin of the *Da* climates in Asia and the United States and are responsible for wide fluctuations in crop yields from year to year. The natural vegetation is chiefly deciduous hardwood forest in the more humid areas and prairie grass in the drier sections. The prairie grass soils are among the most fertile in the world and are thus a major physical asset, accounting to a large degree for the high productivity of the North American Corn Belt.

HUMID CONTINENTAL COOL-SUMMER CLIMATE *(Db)*

This climate is different from the *Da* climate in a number of important respects: The summers are cooler; the frost-free season is shorter; the winters are more severe; snow remains on the ground for a longer period of time; and there is usually less total precipitation. The warm-month temperatures range from 60 to 70° F. in most localities. No month can be above 71.6° F., and at least 4 months must be above 50° F. (Fig. 6–18). The short frost-free season of 90–140 days greatly restricts the variety of commercial crops that can be produced successfully. Cold-month temperatures range from −15 to 26° F. Winters are long, and snow remains on the ground for several months. Cold waves and very low temperatures associated with the importation of cold, dry cP air are not uncommon in the winter season. Precipitation for the year varies from 10 to 40 inches. A higher per cent of the precipitation is in the form of snow than in the *Da* climates. *Dbf* climate prevails in North America and coastal Asia and *Dbw* climate is usually found in the Asian interior.

The *Db* climates are located directly poleward of the *Da* climates in the Northern Hemisphere at latitudes from approximately 45° to as high as 60° in northern Eurasia (Fig. 6–17). In North America, the *Db* climate extends from New England westward beyond the Great Lakes region into the Great Plains of northern United States and the prairie provinces of Canada (Fig. 6–19). *Db* climates are also found in northern China and the most northerly islands of Japan. The greatest extent of this climate, however, is found in central Eurasia, extending from central and northern Europe inland to central Asia. This roughly triangular area is often referred to as the *fertile triangle* and is the only large area within the U.S.S.R. that is productive agriculturally. The short growing season limits large-scale agriculture to hardy, fast-maturing grains, and root and hay crops.

Along the Great Lakes, where the influence of water is a dominant factor, the growing seasons are longer and the bitter-cold winters are tempered. As a consequence, fruits and vegetables are produced successfully along a narrow strip adjacent to the Lakes at latitude 45° and beyond. The original vegetation and soils are similar to those found in the *Da* climates. The humid continental climates together (*Da* and *Db*) cover less than 10 per cent of the world's land area and include about 15 per cent of the world's population.

SUBARCTIC CLIMATE *(Dc, Dd)*

The subarctic climate is characterized by severe winters of long duration and short, cool summers. The greatest annual ranges of temperature on earth are found in these climates, as well as the lowest official midlatitude temperatures. Winter temperatures are extreme, and the cold month averages from 15 to −60° F. or lower. If the cold month average is below

FIG. 6-19. Farm in southern Saskatchewan. Wheat and other small grains are associated with humid continental (*Db*) climates in North America and Europe. (Canadian Consulate General, New York.)

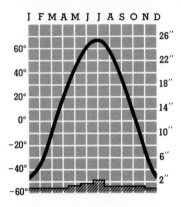

FIG. 6-20. *Dwd* climate—Yakutsk, U.S.S.R. Annual temperature, 13°; range, 112°; annual precipitation, 13.7″. The cold month is —46° and most rainfall is concentrated in the summer.

—36.4° F., the letter *d* is used to denote very severe winters (Table 6–1). Temperatures of —95° F. and lower have been recorded in certain localities during exceptionally cold periods. One to three months in the summer average 65–70° F. (Fig. 6–20), but it is not usual for midday temperatures to reach 85° F. or higher.

The great extremes of temperature are the result of the continental locations and the long nights in the winter season and long periods of daylight during the summer. The frost-free season of 30–90 days is so brief as to preclude widespread commercial agriculture. The long periods of daylight (24 hours at the Arctic Circle at the time of the summer solstice) offset to some degree the briefness of the summer. Total precipitation for the year is usually low, averaging 10–25 inches. The cool summers and long, cold winters result in low evaporation rates, which compensate for the meager total annual precipitation. Precipitation is concentrated in the summer months and is chiefly cyclonic during both seasons.

The subarctic climates are found from latitudes 50° to 70° over the North American and Eurasian land masses (Fig. 6–17). They cover at least one-tenth of the land areas of the world. The cool, short summers, with frost-free seasons of 90 days or less, prohibit the development of large-scale agriculture. Bitter-cold, unfertile soils, often poorly drained as the result of *permafrost* (permanently frozen subsoil), further handicap extensive agricultural development.

The subarctic areas of North America and Eurasia are largely covered by coniferous forests (Fig. 6–21). The greatest contiguous forest area of the world is found in subarctic Eurasia. The trees are relatively small, however, and do not grow as densely as in tropical or subtropical forests. The inaccessibility of these coniferous forests is a major obstacle to their widespread utilization in the foreseeable future. Although the subarctic climate covers a significant part of the earth's land area, only a fraction of 1 per cent of the world's peoples live in these regions.

POLAR CLIMATES (*E*)

The polar climates are less widespread than any other climatic type. They are characteristically found at high latitudes over relatively large land and water areas but, like other cold climates, may be found locally at any latitude if the elevation is sufficient. The polar climates have one outstanding characteristic—the lack of a warm season. In other ways they exhibit a variety of conditions. Their diversity of characteristics and areal distribution reflects the influence of other physical controls in addition to latitude. Although these climates are extremely sparsely settled, their strategic location and their influence

FIG. 6-21. Taiga, or coniferous forest, British Columbia. Spruce, pine, and larch are typical plant associations found in the subarctic (*Dc, Dd*) climates in North America and Eurasia. (Canadian Consulate General, New York.)

over world weather and long-range climatic changes warrant study comparable to that given to the other climatic types.

Polar climates by definition have all months of the year with average temperatures of below 50° F. The 50-degree isotherm for the warmest month serves as the equatorward boundary of the *E* climates and also

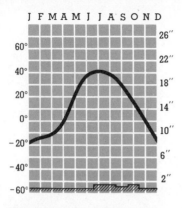

J F M A M J J A S O N D

FIG. 6-22. *ET* climate—Barrow Point, Alaska. Annual temperature, 10°; range, 59°; annual precipitation, 5.6". The warm month is between 32° and 50°.

approximates the poleward limit of forest growth. The *E* climates are divided into two subgroups: (1) the tundra climate (*ET*), in which the warm month averages between 32 and 50° F.; and (2) the ice cap climate (*EF*), where the warm month is below 32° F.

TUNDRA CLIMATE (*ET*)

A long, cold winter and a few frost-free days in the summer are the marks of the tundra climate. In the summer, the snow or ice cover disappears as a result of the long periods of daylight, and the upper few feet of ground thaw. The subsoil, however, remains permanently frozen, impeding drainage; and wet, swampy surface conditions prevail. Warm-month temperatures average between 32 and 50° F., and frosts are common during the brief summer (Fig. 6–22). Coastal locations usually experience less severe cold than interior areas. Precipitation is meager—10–15 inches as a general average, but some coastal areas may receive considerably more. It is cyclonic in origin and is often in the form of snow. The warmer

FIG. 6-23. The tundra, found poleward of the tree line. Vegetation consists of hardy grasses, mosses, sedges, and lichens, and is associated with the tundra (*ET*) climate. (Canadian Consulate General, New York.)

EARTH SCIENCE

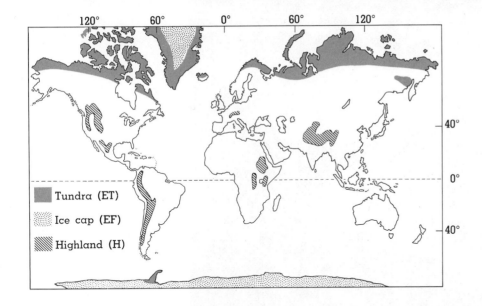

FIG. 6-24. Generalized map showing distribution of the polar and highland climates: tundra (*ET*), ice cap (*EF*), and highland (*H*).

Legend:
- Tundra (ET)
- Ice cap (EF)
- Highland (H)

air during the brief summer holds more moisture and, consequently, most precipitation occurs during this season.

The *ET* climates are poleward of the tree line. The natural vegetation, referred to as tundra, consists of mosses, sedges, lichens, hardy grasses, and dwarf, bushlike trees (Fig. 6–23). The exact type of tundra varies with local climatic and soil conditions, but in all cases it is adapted to dryness, for the moisture is most frequently frozen as ice or snow and is thus unavailable to plants. The tundra serves as forage for reindeer and caribou and is the basis of a migrating type of pastoral economy carried on by the Lapps and other nomads of Eurasia.

The *ET* climates are located at latitudes 55° to 80° along the poleward fringes of the Eurasian and North American continents. They are also found along the margins of the ice caps in Antarctica and Greenland (Fig. 6–24). Only a few hundred thousand people occupy the tundra regions— hunters, miners, nomads following reindeer herds, and military personnel and scientists manning defense or scientific outposts.

ICE CAP CLIMATE *(EF)*

The ice caps are the most inhospitable regions of the earth in terms of resource potentialities for human life (Fig. 6–25). They are occupied on a temporary basis only, by a few scientists, hunters, and military personnel. Few data have been compiled concerning weather conditions, and

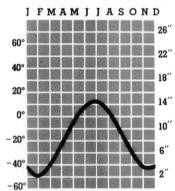

FIG. 6-26. *EF* climate—Eismitte, Greenland. Annual temperature, —22°; range, 65°; annual precipitation, no data. The warm month is below 32°.

the ice caps are the least known of the climatic types. It is certain, however, that average annual temperatures are the lowest of any climate, and severe cold prevails most of the year. The cold months average —30 to —60° F., the warm months usually 0 to 32° F. (Fig. 6–26).

There is little information concerning precipitation, but it is probably always in the form of snow, with total annual amounts equivalent to 1–5 inches of rainfall. Precipitation is cyclonic in origin and is likely to be heaviest in the warmer part of the year. The ice caps cover most of Greenland and Antarctica, and small areas of permanently frozen ice in the vicinity of the North Pole (Fig. 6–24).

The *E* climates cover about one-tenth of the world's land area and probably have a permanent population of not more than ¼ million people.

HIGHLAND CLIMATE (*H*)

The highland climate is not as clearly defined as the other climatic types. But since highlands of 3000 feet or more are markedly different in climatic characteristics from their adjacent lowlands, it is useful to group them in a separate category (Fig. 6–24). Specific weather and climate conditions in these regions depend upon altitude, latitude, location on windward or leeward slopes, and exposure to sunlight and winds (Fig. 6–27 shows data for one *H* climate locality). But *H* climates all differ from adjacent lowlands in the following respects:

1. Air pressure is drastically lower. We have noted (Chapter 3) that air pressure decreases about 1 inch per 1000 feet of increase in elevation at lower altitudes (Table 3–1).

2. Air temperature is markedly lower. The temperature decreases at an average rate of 3½° per 1000 feet of increase in elevation. The vertical decrease in temperature is many hundredfold greater than the horizontal gradient. At the equator, the gamut of climates from *A* to *E*, can be experienced along a mountain slope that extends from low altitudes to about 15,000 feet. At a uniform altitude, however, several thousands of miles from the equator to polar latitudes would have to be traversed in order to experience the same climatic transitions. The vertical zoning of climates along highlands produces a very conspicuous corresponding zoning of natural vegetation and crops (Fig. 6–28).

3. Solar energy is more intense. With increased elevation there is generally a decrease in the water vapor content of the air as well as in the quantity of dust particles. These are the two principal elements that absorb, reflect, and scatter solar radiation, and the intensity of sunlight therefore increases as they decrease with elevation. The increase is greatest at elevations of 6000 feet and over, for on a clear day more than three-quarters of the solar energy penetrating the atmosphere reaches 6000 feet, whereas less than one-half penetrates to sea level. The greater intensity of sunlight has an important influence on soil temperatures, plant growth, and sensible temperatures.

4. Daily ranges of temperature are considerably greater. The drier and cleaner air of highland areas results not only in stronger solar radiation during the day but also in the rapid loss of terrestrial energy at night. The diurnal or daily range of temperature is thus much greater than in adjacent lowlands. Annual ranges of temperature, however, are essentially the same, varying in accordance with the latitude and the location with respect to continents and ocean bodies.

5. In highland areas along windward slopes precipitation is heavier. Not only is rainfall greater, but evaporation is less, owing to the lower temperatures. The humid windward slopes are important to their adjacent lowlands, for they are the source of streams that are used for irrigation, power, and in some cases transportation. The leeward slopes are usually drier, for they are within the rain shadow of the highlands.

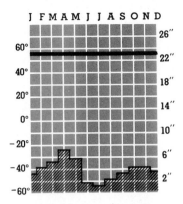

FIG. 6-27. *H* climate—Quito, Ecuador. Annual temperature, 54°; range 0.5°; annual precipitation, 42.3". Cool temperatures for the latitude; little annual range of temperature.

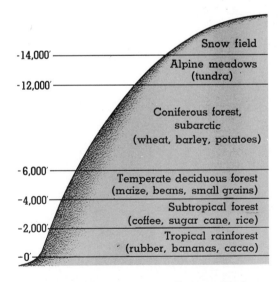

FIG. 6-28. Highly generalized diagram showing vertical temperature zones as they might exist in the tropical latitudes.

155

6. Winds are stronger on exposed summits and slopes, where the effects of surface friction are minimized. Highland areas are subject to a number of local winds and accompanying weather conditions such as mountain and valley breezes and *chinook,* or *foehn,* winds.

Weather shifts within the period of a day tend to be greater and to take place more rapidly in the highlands than in adjoining lowlands. Rapid temperature changes, winds varying in direction and in velocity from calm to gale proportions, and rain and snow changing to intense sunshine are likely to make up the daily weather, in contrast to the more stable conditions of the lowlands.

OTHER CLIMATIC CLASSIFICATIONS

The Köppen classification of climates is the one most widely used. Other systems have been developed, however, and the Thornthwaite classification deserves special mention. In 1931, C. Warren Thornthwaite introduced a classification that was applied to North America; in 1933 it was extended to cover the world. Like the Köppen classification, it is a quantitative system, utilizes a combination of symbols to designate types of climate, and is based essentially on the character of vegetation found associated with each major climate.

Unlike the Köppen system, it utilizes *precipitation effectiveness* and *temperature efficiency* rather than monthly and annual precipitation and temperature data. Precipitation effectiveness (P/E ratio) is calculated by dividing the total monthly precipitation by the total monthly evaporation. The sum of the twelve monthly P/E ratios is called the P-E index. Because of the lack of data on evaporation, Thornthwaite devised a formula based on temperature and precipitation relationships to determine the P-E index: P-E index = sum of twelve monthly values of $[115(P/T - 10)(10/9)]$, where T is mean monthly temperature in °F and P is mean monthly precipitation in inches.

In a similar manner, the sum of the twelve monthly temperature efficiency ratios (T/E ratios) indicates the T-E index: T-E index = sum of twelve monthly values of $T - 32/4$.

On the basis of the P-E index, five humidity categories are defined— rain forest, forest, grassland, steppe, and desert—which correspond to major vegetation types. The T-E index designates categories as tropical, mesothermal, microthermal, taiga, tundra, and frost. Further subdivisions are determined on the basis of seasonal distribution of precipitation.

In theory, 120 combinations are possible utilizing the three criteria P-E index, T-E index, and seasonal distribution of rainfall. In actuality, however, 32 climatic types are designated on the world map of climates that uses this system.

In 1948, Thornthwaite proposed a climatic classification based on the concept of *potential evapotranspiration*. The same three criteria were employed as in his earlier system: precipitation effectiveness, temperature efficiency, and seasonal distribution of precipitation. Evapotranspiration refers to the transfer of moisture from ground to atmosphere by the combined mechanisms of evaporation from the soil and transpiration by plants. The comparison of the amount of moisture input from precipitation with the potential evapotranspiration, or the maximum amount of moisture output if it were available, provides a system for determining areas of water surpluses and deficiencies. A chart illustrating the annual march of monthly potential evapotranspiration and monthly precipitation indicates the balance of moisture, both annual and seasonal.

Since no actual measurements of potential evapotranspiration are available, Thornthwaite was compelled to compute this quantity as a function of temperature. On the basis of the computed indexes of moisture and heat, maps of the United States showing average potential evapotranspiration, average annual temperature efficiency, seasonal effective moisture, and summer concentration of temperature efficiency have been constructed. No map of world climates using this classification have been produced, primarily owing to the lack of potential evapotranspiration data for large areas of the world. However, maps of annual water surpluses and deficiencies for various parts of the earth have been developed.

SUGGESTED REFERENCES ————————————————

Blair, T. A.: *Climatology, General and Regional*, Prentice-Hall, Inc., Englewood Cliffs, N.J., 1942.

Climate and Man, U.S. Department of Agriculture, 1941 Yearbook, Government Printing Office, Washington, D.C., 1941.

Critchfield, H. J.: *General Climatology*, Prentice-Hall, Inc., Englewood Cliffs, N.J., 1966.

Geiger, R.: *The Climate Near the Ground*, Harvard University Press, Cambridge, Mass., 1957.

Kendrew, W. C.: *The Climates of the Continents*, Oxford University Press, New York, 1961.

Koeppe, C. E., and G. C. DeLong: *Weather and Climate*, McGraw-Hill Book Company, Inc., New York, 1958.

Sellers, W. D.: *Physical Climatology*, University of Chicago Press, Chicago, 1965.

Trewartha, G. T.: *An Introduction to Climate*, McGraw-Hill Book Company, Inc., New York, 1968.

Visher, S. F.: *Climatic Atlas of the United States*, Harvard University Press, Cambridge, Mass., 1954.

THE SOLID EARTH: MINERALS AND ROCKS

As yet, we probably know more about what happens inside the most distant stars than we do about the structure of our own planet. The methods by which we learn about the stars are of no avail to us in studying the solid earth. As noted earlier, scientists have been unable to agree on a single, unquestionably correct theory of the origin of the solar system. Nor have they been able to establish beyond doubt a fully acceptable theory of the earth's beginnings. If we knew how the earth began, we could develop at least a working hypothesis about its subsequent evolution. But at the present time we can still do little more than speculate—that the earth may have originated as an incandescent body of gas that slowly cooled, first condensing to a liquid and then solidifying, in which case it is still cooling and contracting; or, to the contrary, that it may have originated by the slow accumulation of solid fragments which heated subsequently.

Our best source of information on the earth's interior is the behavior of earthquake waves. From detailed study of their speed and direction as they travel within the earth, we can make assumptions about the substances through which they are passing. Thus we have formed a tentative picture of the earth's structure:

A central *core*, of radius of about 2200 miles, which is in a liquid state but probably has an innermost zone that is solid

A surrounding *mantle* about 1800 miles thick, composed of relatively dense, rigid rock

The *crust*, ranging in thickness from about 8 miles beneath the oceans to 25–40 miles in continental areas

Most of our knowledge concerns the crust, since it includes the surface and materials that have reached the surface from the depths. Reasonably good evidence indicates that it consists of an outer, relatively thin layer of sedimentary, or deposited, material; a layer generally extending 6 to 9 miles under the continents composed predominantly of granite and granitelike rock; and a layer of basaltic composition under both continents and oceans with its base at depths ranging from 8 to 40 miles.

We shall return to this picture in a later chapter, to examine it in greater detail. But first we need to know something of the materials and structures that make up the solid earth. What is granite? What is basalt? Of what are rocks composed?

MINERALS

Insofar as we can judge from materials observed at the earth's surface and to depths of almost 5 miles in mines and oil wells, the principal components of the solid earth are discrete crystalline grains of minerals. They make up the rocks and a large percentage of the soil, which is composed for the most part of weathered rock material. A mineral may be defined as a naturally occurring inorganic substance with a definite chemical composition, a regular crystal structure, and external form and physical and chemical properties that are related to its composition and crystal structure. Although much of this definition would fit many man-made materials, our interest is only in the naturally occurring substances formed by geologic processes and we shall restrict the term to these.

Because each mineral has a definite chemical composition, each can be identified by a chemical formula as well as by its mineralogic name. For example, the mineral commonly known as rock salt is called halite, and its chemical formula is $NaCl$, denoting an equal number of atoms of sodium (Na) and chlorine (Cl). All halite, no matter whether from Germany, New Mexico, or Michigan, has the same chemical composition—although trace amounts of impurities may be present, since natural conditions are not often as controlled as those in the laboratory. It is interesting to note that the name of a mineral may or may not give us useful information about it, whereas its chemical formula states its constituents precisely. "Hematite" comes from the Greek word meaning "bloodlike," and denotes the blood-red color of the mineral in powdered form; "quartz," on the other hand, is of unknown origin and is simply a commonly accepted label. But if we know the formula for quartz—SiO_2—and a little elementary chemistry, we know the most essential information about it: that it is composed of silicon and oxygen atoms in a ratio of $1:2$.

There are some naturally occurring substances which are of organic, rather than inorganic, origin. These are mixtures of many components, and do not have the same composition in all occurrences nor even from sample to sample. One example is coal, which is formed from preserved and altered plant tissues comprising hundreds of different chemical compounds. Such a substance cannot be called a mineral. Do not be confused, however, when you find that coal is grouped with the true minerals in discussions of mineral resources.

In general, a specific chemical composition is accompanied by an orderly internal arrangement. Halite ($NaCl$), for example, has sodium atoms and chlorine atoms alternating in a cubical arrangement, with unit after unit added to make the specimen we see in the laboratory or the deposit we see in the mine. As a result of this structure, halite takes the external form of cubic crystals, and when a specimen is broken by a sharp blow with a hammer, the pieces have square corners. Any regular pattern such as we have described is called a crystalline structure, and it is characteristic of nearly all minerals, as well as many artificial materials. The particular form of crystalline structure is the geometric result of the uniform size of all atoms of each component element, the relative numbers of the different kinds of atoms, and the arrangement of atoms. Naturally occurring substances that are mixtures are not and cannot be crystalline because the mixtures differ from sample to sample and the components are not organized in an orderly arrangement.

The chemical and physical properties of a mineral are related to its chemical composition and molecular structure. Hence any specimen of a particular mineral will have the same properties as any other no matter where on earth it comes from, or, in fact, even if it is found in a meteorite from space. A specimen of the iron ore hematite from Michigan is identical in its chemical and physical properties to hematite mined in Minnesota, Wisconsin, Labrador, Brazil, India, and other places, and could not be identified as to its place of origin. We are fortunate, in one respect, that this is the case, for otherwise the processes of mining, concentration, and purification of ores would have to be different for each deposit.

CHEMICAL COMPOSITION AND MOLECULAR STRUCTURE

We have seen, then, that the properties of a mineral depend upon its atomic makeup. The behavior of minerals and the rocks they comprise under various forms of environmental stress is also dependent upon this basic structure. It will be useful, therefore, to give some brief consideration to a few key concepts required for an understanding of the composition of matter.

All matter, living and nonliving, is composed of atoms; consequently the behavior of matter in any form and under all conditions may be fundamentally understood in terms of the forces of attraction and repulsion within and between atoms. It would be far too cumbersome, and not essential to our comprehension of large-scale phenomena, to attempt to analyze every event in terms of its atomic processes. Nevertheless, we should be aware that whatever aspect of nature we discuss is built upon the complex interactions of atoms, molecules, elements, and compounds.

Elements. An *element* is defined as a substance which cannot be decomposed into different substances either by physical processes not affecting the atomic nucleus or by chemical means. Halite ($NaCl$) can be separated into the metal sodium (Na) and the gas chlorine (Cl); thus it is not an element but a compound. Sodium and chlorine cannot be separated, so they are elements.

At present, 103 elements are known—most of them natural, some artificial. Each element has its special arrangement of the fundamental particles of matter, with a specific number of protons and electrons. (Some elements have varying numbers of neutrons; these forms of the same element are called *isotopes*.) The atom is the smallest subdivision that

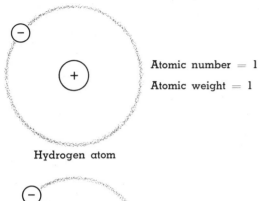

Atomic number = 1

Atomic weight = 1

Hydrogen atom

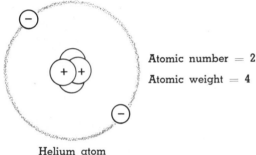

Atomic number = 2

Atomic weight = 4

Helium atom

FIG. 7-1. Models of hydrogen and helium atoms.

retains the characteristics of the element. Physicists have discovered many kinds of so-called elementary particles and continue to discover more. But for our purposes we shall restrict our analysis of the atom to three subatomic particles—*protons, electrons,* and *neutrons.* Note that the arrangement of these particles within, say, a sodium atom makes a sodium atom different from an atom of any other element, but the protons, electrons, and neutrons are not distinguishable from one kind of atom to another.

Our picture of the atom is based on concepts first proposed by the Danish physicist Niels Bohr in 1913.

Atomic particles. The electron is a particle having a negative electric charge, and the proton has an equivalent positive charge. The neutron is electrically neutral. Each atom contains an equal number of protons and electrons and hence equal negative and positive charges. Thus the atom as a whole is neutral. The proton is far heavier than the electron—its mass is 1837 times greater. The neutron is a particle of almost the same mass as the proton. The atoms of the different elements are composed of protons and neutrons in different numbers in a dense nucleus surrounded by electrons in orbital motion. Electrons move about the nucleus three-dimensionally in a series of orbits whose configurations are too complex for us to consider here. The term *shells* is used to denote the positions, at all possible times, of electrons moving at specific distances from the nucleus.

Let us look now at models of some atoms. The hydrogen atom, the simplest known, consists of a nucleus containing one proton, and one electron revolving around the nucleus and at a distance from it (Fig. 7–1). The negative charge of the electron balances the positive charge of the nucleus. The next larger atom, helium, consists of a nucleus containing two protons and two neutrons, and two electrons revolving about the nucleus (Fig. 7–1). Again, negative and positive charges balance. The elements are given *atomic numbers* denoting the number of protons in the nucleus; thus hydrogen has an atomic number of one and helium of two. The *atomic weight* is the total of protons and neutrons in the nucleus, so hydrogen and helium have atomic weights of one and four, respectively.

All the information we have indicates that the arrangement of electrons in relation to the nucleus is limited to certain definite patterns, some of which are stable and some of which are not. The particular pattern which prevails and its stability or instability determine the chemical characteristics of an element. For example, the hydrogen atom, with one electron, is unstable; hence hydrogen is chemically very active. It combines with oxygen so readily that it is considered an explosive material and is therefore ordinarily not used to inflate balloons or blimps. Helium, on the other hand, with two electrons, is a most stable material and under ordinary conditions is completely inactive. The United States uses helium in preference to hydrogen in lighter-than-air craft.

Electrons move around atomic nuclei in a series of concentric shells, each at a specific distance from the nucleus. The single electron of hydrogen and the two electrons of helium are in the inner shell. Larger and more complex atoms have two electrons in the inner shell (the maximum for this position) and additional electrons in shells at greater distances from the nucleus. Any element that has eight electrons in its outermost shell is stable and thus does not combine chemically in nature with other elements (Fig. 7–2). There are five such elements; these, together with helium, are known as the *noble gases*. The other elements can combine chemically.

Compounds. Compounds are combinations of elements. Elements having an outer shell with fewer than eight electrons can combine chemically to form compounds. An element with only one electron in its outer shell easily loses the single electron, and an element with seven electrons in its outer shell has a tendency to pick up an electron to make a complete shell of eight. Let us consider, for example, halite, made up of sodium (Na) and chlorine (Cl). Sodium has only one electron in its outer shell and chlorine has seven. The sodium atoms each lose one negatively charged electron, thus becoming positively charged, and the chlorine atoms pick up single electrons and become negatively charged (Fig. 7–3). The oppositely charged particles attract each other and combine to form the compound sodium chloride (NaCl). The sketches show the positive sodium ion and the negative chloride ion. Arrows show electrons easily lost or gaps easily filled. Atoms that have become positive or negative by losing or gaining electrons are called *ions* (recall that atoms are neutral, since they have an equal number of positive and negative particles). Halite is called an

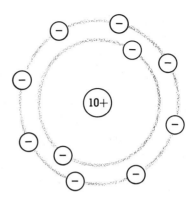

FIG. 7-2. A stable element: neon, containing eight electrons in outer shell.

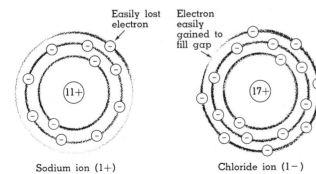

Sodium ion (1+)   Chloride ion (1−)   **FIG. 7-3. Sodium and chloride ions.**

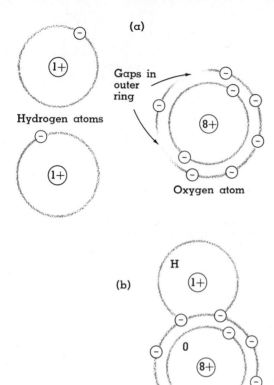

(a)

Hydrogen atoms

Gaps in
outer
ring

Oxygen atom

(b)

H

O

H

Water
molecule

FIG. 7-4. Water (H₂O), a compound composed of hydrogen and oxygen. (a) Two hydrogen atoms and one oxygen atom. (b) One water molecule, showing the shared electrons.

ionic compound because it is composed of ions. Its characteristics are entirely unlike those of either of its components, sodium and chlorine.

Many compounds are composed not of ions but of molecules—atoms of different elements sharing electrons. Ordinary water (H_2O) is a compound of hydrogen, with one electron in its single shell, and oxygen, with six electrons in its outer shell. Water molecules consist of one oxygen atom and two hydrogen atoms, sharing electrons and thus completing the single, inner shell of hydrogen with two, and the outer shell of oxygen with eight, as indicated in Fig. 7-4. A water molecule can be divided into atoms of hydrogen and oxygen, but it cannot be divided into parts which have the properties of water. Thus the molecule is the smallest unit of a compound which has the properties of the substance. Minerals may be elements or they may be compounds made up of elements combined in the form of either molecules or ions.

The common elements. By analyzing rocks and computing their volume, scientists have found that eight elements are by far the most common components of the earth's crust, as shown in Table 7-1. The next two most abundant elements make up approximately another ¾ of 1 per cent, so you can see that 10 of the 103 known elements account for virtually all of the crust of the earth. Some elements we think of as abundant, such as copper, are really very rare in comparison with those that are listed.

IDENTIFICATION OF MINERALS

To learn to identify all the known minerals would be an enormous task, for there are about 2500 (although some of these are so similar

TABLE 7-1   MOST COMMON ELEMENTS IN THE EARTH'S CRUST[a]

ELEMENT	WEIGHT PERCENTAGE OF THE EARTH'S CRUST
Oxygen (O)	46.60
Silicon (Si)	27.72
Aluminum (Al)	8.13
Iron (Fe)	5.00
Calcium (Ca)	3.63
Sodium (Na)	2.83
Potassium (K)	2.70
Magnesium (Mg)	2.09
Total	98.70

[a] Jack Green, Geochemical Table of the Elements for 1953, *Bulletin of the Geological Society of America*, vol. 64, no. 9, Sept., 1953.

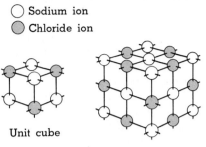

○ Sodium ion
● Chloride ion

Unit cube

Augmented unit cube

Larger cubic crystal

FIG. 7-5. Growth of halite crystal by addition of unit cells.

that there is controversy about listing them separately). The large number is not surprising when you think of the many possible combinations of elements and the many different environments in which minerals can form. Fortunately, 80 to 100 of the most common minerals account for 99 per cent of the total mineral volume, and to identify the most common rocks, 20 or 30 of the most common minerals suffice. The differences among these are so distinct that one can become competent in differentiating them in a comparatively short time. The physical properties that are most useful in identifying minerals are related either to crystal structure or to chemical composition, or they are optical—that is, having to do with light.

PROPERTIES RELATED TO CRYSTAL STRUCTURE

Crystal form is governed by internal crystal structure. The crystalline substance halite (NaCl) has a crystal unit (the unit cell) which is a cube. Addition of more and more cubes to increase the size of the grain may result in a crystal which is a cube. This method of growth is shown in Fig. 7–5. It is also possible for a cubic crystal to be enlarged more in one direction than in another; in that case the cubic form is distorted to that of a brick. In either event, however, the angles between the corresponding faces do not change, and this property may often be more useful in identification than the crystal form itself. A cubic crystal structure may also be the basis for an external form different from a cube; for example, a double-ended pyramid, shown in Fig. 7–6. Either or both of these external crystal forms may occur in a given mineral having a cubic crystal struc-

FIG. 7-6. Growth of a pyramid by addition of cubic units.

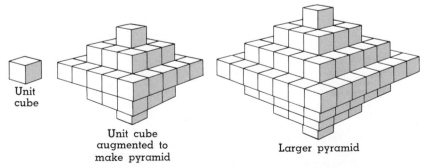

Unit cube

Unit cube augmented to make pyramid

Larger pyramid

FIG. 7-7. Feldspar crystal. Flat surfaces are the exterior form. (Ward's Natural Science Establishment, Inc.)

ture. Some cubic minerals preferentially crystallize in one of the two forms but others do not. There is a wide range of crystal forms (Fig. 7-7).

Similar relationships exist between crystal structure and external crystal form in most minerals. Where other properties of two mineral specimens are alike (or perhaps impossible to determine) the crystal form may be the deciding factor in making an identification.

Cleavage is the preferential breaking of a crystalline material along some direction related to the crystal structure to form smooth plane surfaces. Halite (NaCl) is an excellent example (Fig. 7-8). If we bear in mind the cubic structure, it is reasonable to expect that when a fragment of halite crystal breaks, it will break along planes parallel to the faces of the cube, the planes of weakness being parallel to the faces. This is, in fact, what occurs. We say, therefore, that halite has *cubic cleavage*, the planes being at right angles to each other. Another familiar cleavage is that of the micas. They can be roughly described as having a hexagonal pattern of closely spaced units in a sheet, and sheets which are relatively widely spaced. The sheets can be easily separated, and the result is cleavage in one direction.

Other cleavages are two at right angles, two at oblique angles, and three, four, or six cleavages at various angles, some at 90° and some oblique (Fig. 7-9). There may be difficulty in distinguishing between the smooth, reflecting external surfaces of the crystal form and the cleavage, or internal feature. Cleavage is indicated by the numerous repeated parallel surfaces, steplike in appearance, on a broken specimen. It is necessary to look on the specimen for a broken surface to be sure that the mineral has cleavage and to determine the number of cleavage directions and the

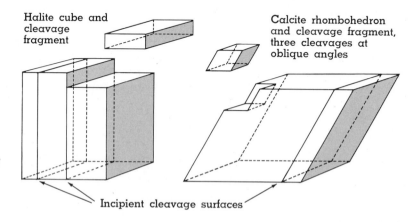

Halite cube and cleavage fragment

Calcite rhombohedron and cleavage fragment, three cleavages at oblique angles

Incipient cleavage surfaces

FIG. 7-8. Mineral fragments, showing cleavage. All external surfaces are cleavage planes. Note parallelism of sides and edges. Incipient cleavage surfaces are indicated.

angular relationships between them. If there are no planes of weakness in a mineral, it will not cleave even though there is a regular crystal structure.

PROPERTIES RELATED TO CHEMICAL COMPOSITION

Specific gravity is the ratio of the mass of a specimen of a substance to the mass of an equal volume of pure water. A cubic foot of water weighs 62.4 pounds; a cubic foot of aluminum weighs 168.5 pounds, about 2.7 times as much; hence we say that the specific gravity of aluminum is 2.7. The average specific gravity of the materials making up the continental areas of the earth's crust is close to 2.7. By "hefting" a specimen in one's hand, one can judge, from experience, whether it weighs more or less than the average for its size and shape.

Specific gravity is related both to composition and to crystal structure. Minerals containing lead, a heavy element, have a higher than average specific gravity. And a substance in which the atoms are packed closely together has a higher specific gravity than one in which the atoms are far apart.

Hardness is the ability of a mineral to scratch other minerals. There is a great range from the softest to the hardest, and the degree of hardness of a specimen is very useful in mineral identification. An early German mineralogist, Friedrich Mohs, set up a scale which is used as a standard of comparison. Ten minerals are listed, in order of increasing hardness:

1. Talc	5. Apatite	9. Corundum
2. Gypsum	6. Orthoclase	10. Diamond
3. Calcite	7. Quartz	
4. Fluorite	8. Topaz	

FIG. 7-9. Cleavage fragments of calcite showing three directions of cleavage at oblique angles. (Ward's Natural Science Establishment, Inc.)

Except for the diamond all are common in the pure form and are easily obtained. The difference in hardness between each pair of consecutive minerals is approximately the same except for the great difference between corundum and diamond.

We may test for hardness by attempting to scratch the unknown specimen with the minerals in the hardness set, thus determining a hardness lying between two standard values. In performing the test it is essential to confirm one's judgment by attempting to scratch the standard with the unknown, and also by rubbing to be sure there is an actual scratch rather than merely a chalk mark on the surface.

An incomplete but useful hardness set can be made up of these everyday materials:

Hardness	Material
2½	Fingernail (there are some exceptions)
3	Copper penny
5	Knife blade
5½–6	Window glass (not Pyrex or optical glass)
6½–7	Steel file

The hardness of a mineral may be related to its chemical composition, to its crystal structure, or to the kind of forces holding the atoms together in its molecules.

PROPERTIES RELATED TO LIGHT

Luster is the quality of light reflected from the surface of a mineral. Metallic luster is the appearance of clean, untarnished metal. The high degree of reflectivity and distinctive appearance of bare metal are readily recognizable regardless of color. The nonmetallic lusters are divided into a number of subcategories. Adamantine luster is the piercing brilliance of diamond, and it is uncommon. It grades into glassy (or vitreous) luster, the luster of a piece of clean glass. Related is greasy luster, the appearance of glass covered by a thin film of oil. Waxy luster is like that of paraffin, beeswax, or candles—i.e., some light coming from within a cloudy material. Silky luster is that of a bundle of fine parallel fibers, such as silk, nylon, or mercerized cotton. Pearly luster has an iridescent color play like mother-of-pearl or pearls. Resinous luster is unique in being associated with a specific color—the yellow-to-amber color of resin. It is seen in few minerals. Earthy luster is that of a dull surface having the appearance of a clod of dirt. A few minerals are of submetallic luster—between metallic and nonmetallic.

Streak is the color of a mineral in its powdered form. Since it is not always possible or practicable to crush a specimen to a fine powder, the normal method of ascertaining streak is to draw an edge or corner of the mineral along the surface of an unglazed white porcelain plate and to examine the color of the chalk mark left on the plate. The color contrast is good against the white plate, but it is even better if the ball of the thumb is rubbed lightly across the streak to smear it out a little. Several short streaks made by different parts of the specimen are much better than one long, broad swipe, which tells no more than a short one and uses up a large part of the clean surface of the plate. Streak is a very reliable property. Some minerals have different appearances in different specimens but almost always have the same streak.

Color of minerals is usually more deceptive than helpful. Many minerals are colorless in the pure form and take on various colors depending on the kind and amount of impurity present or the surface alteration. Beryl, for example, is the source of beryllium, a metal used to make hard, light-weight alloys. It sells for a few cents a pound on the open market. With the right amount of contamination, it has the color of the emerald, a precious stone.

Diaphaneity is the degree to which light passes through a substance. Opaque substances allow no light to pass. In translucent materials light passes through but no images can be seen; and transparent materials allow images to be distinguished.

OTHER PROPERTIES

Fracture is the appearance of a break that is not a cleavage. The fracture surface is neither plane nor is it smooth, as is a cleavage plane. Several types are seen. *Conchoidal* fracture is the curved, semismooth break of glass. *Hackly* fracture is a rough surface with many tiny projections like the break of a metal such as cast iron. *Splintery* fracture is a surface showing a strong parallel, linear appearance similar to a split in a piece of wood. Other kinds of fracture less distinctive in appearance are called rough or irregular. Some minerals have both cleavage and fracture. A mineral which has two cleavage directions at right angles to each other may break into elongated pieces with sides which are cleavage surfaces but will break across the end in a fracture.

Magnetism is a property of a few minerals containing a high percentage of iron—magnetite, for example. They show a strong attraction when exposed to a magnet.

Effervescence when in contact with acid is a characteristic of a few minerals such as calcite. The rate of bubbling, or fizzing, the temperature required, and the need for powdering the mineral are useful in differentiating among these minerals.

Taste is useful in identifying minerals soluble in water, as halite (NaCl), for example.

Feel is useful only rarely. Talc has the smooth feel of dry soap.

EQUIPMENT FOR MINERAL IDENTIFICATION

The most important tool for examining mineral specimens is the hand lens, a pocket magnifying glass with magnification of 7–12 times. It is used to study cleavage, crystal form, scratches made in the hardness test, and streak. A pocket knife is useful to test for hardness, although an ordinary iron carpenter's nail will suffice. The streak plate is used for streaking the minerals, and a magnet to separate the magnetic minerals. Dilute hydrochloric acid in a small bottle with a dropper cap is necessary to test for calcite. This test should be used only if essential, however, in order to avoid spattering acid indiscriminately over all the specimens.

ROCK-FORMING MINERALS

A small group of minerals, among all those found in nature, appear in the most common rocks. It is especially useful to become familiar with their names and their identifying properties. Table 7–2 shows 15 of the most important rock-forming minerals, and a few others that are of economic value, with their outstanding characteristics.

TABLE 7–2   MINERAL IDENTIFICATION KEY

HARDNESS GREATER THAN GLASS				
HAS CLEAVAGE		LACKS CLEAVAGE		
Metallic luster	Glassy luster	Metallic luster	Glassy luster	Dull luster
HEMATITE 1 cleavage, red-brown streak	FELDSPAR 2 cleavages at right angles HORNBLENDE 2 cleavages at oblique angles, greenish black	PYRITE Brass-yellow; black to greenish-black streak MAGNETITE Dark gray to black; black streak, magnetic	QUARTZ Colorless and clear to light colored; light streak GARNET Dark shades of brown and red, rarely green OLIVINE Green, hardness slightly less than quartz	HEMATITE Red, brown to nearly black; red-brown streak CHERT (CHALCEDONY) Light colors; colorless streak, opaque except translucent on very thin edges, hardness 6–7

HARDNESS LESS THAN GLASS

HAS CLEAVAGE		LACKS CLEAVAGE		
Metallic luster	*Glassy luster*	*Metallic luster*	*Glassy luster*	*Dull luster*
GRAPHITE 1 cleavage, gray; gray streak, hardness less than fingernail, conspicuously low specific gravity, greasy feel	**TALC** 1 cleavage, hardness less than fingernail, greasy feel	**COPPER** Copper-colored surface and streak, ductile, conspicuously high specific gravity	**GYPSUM** Light colors, hardness 2	**TALC** Greasy feel, hardness 1
HEMATITE 1 cleavage, red-brown streak	**MUSCOVITE** 1 cleavage, sheets flexible and elastic, colorless		**SULPHUR** Yellow, hardness 1½–2½, burns	**GYPSUM** Light colors, hardness 2
GALENA 3 cleavages at right angles, dark-gray streak, hardness near fingernail, conspicuously high specific gravity	**BIOTITE** 1 cleavage, sheets flexible and elastic, dark green to dark brown or black			**KAOLINITE** Light colors, smells earthy when breathed on
	CHLORITE 1 cleavage, sheets flexible but inelastic, green			**LIMONITE** Shades of brown; yellow-brown streak, variable hardness
	GYPSUM 1 cleavage prominent, 3 others inconspicuous, sheets flexible but inelastic, light colors usually, hardness 2			**HEMATITE** Shades of brown and reddish brown to nearly black, red-brown streak
	FELDSPAR 2 cleavages at right angles			
	HORNBLENDE 2 cleavages at oblique angles, dark green			
	HALITE 3 cleavages at right angles, hardness 2–3, salty taste			
	CALCITE 3 cleavages at oblique angles, fizzes in acid			

You will note from the table that in some cases the same mineral may either have or lack cleavage and may have more than one type of luster. A cleavage surface can be no larger than the mineral grain cleaved, so a mineral in extremely fine granular form may not show cleavage except at high magnification. Luster may likewise differ with grain size. When a mineral has an earthy or dull luster it is likely to be composed of fine grains and may be porous, and it will be softer, of lower specific gravity, and perhaps lighter in color than the same mineral with another type of luster. In such instances the streak will usually remain reliable for identification purposes.

Rocks

If minerals are the principal components of the solid earth, rocks may be said to be its form. What is rock? A rock is an aggregate of particles of one or more substances, generally mineral. (Some rocks are composed of coal, and you will recall that coal is not a mineral.) Most rocks are combinations of several minerals, but not all. (Limestone, for example, is made up of essentially one component, calcite.) Perhaps the most characteristic fact to be noted about rocks is that they may be formed of an almost infinite number of combinations of minerals in different proportions and different physical distributions. To make matters even more complex, rocks never exist in a state of isolation—even though, for the sake of simplicity, we may study them that way—and since that is true, the characteristics of a rock are always in some degree related to its surroundings, the conditions under which it was formed, and the changing conditions under which it exists.

Although we ordinarily think of rock as a hard, strong material, in the geologic sense a deposit of loose sand does qualify as a rock unit. However, for our purposes we shall consider only those rocks that are cemented or in other ways consolidated.

Despite the diversity we have noted, geologists have formulated a few general rock categories, based on the manner of formation, that enable us to bring some order to the study of the rocks that we observe at the earth's surface. Three rock groups are traditionally recognized—*igneous, sedimentary,* and *metamorphic.* For each of these, certain features are characteristic of most specimens. Yet, because of the great range of environments in which rocks may come into being, there is no single, easily recognized characteristic unique to each group and present in all its samples. It is usually necessary, therefore, to recognize a combination of characteristics in order to make a reliable identification of any rock specimen. Even so, field relationships are so complex that rocks may be found which are difficult to name. But laboratory specimens that you will see are, for the most part, excellent examples of their group.

IGNEOUS ROCKS

Igneous rocks are those which form by the cooling and solidifying of a body of molten rock material, either at depths within the earth or at the surface. It is thought that rocks of this type make up about 95 per cent of the outermost 10 miles of the earth. Generally they are hard, dense, and lacking in pore space, or voids.

The molten material from which igneous rocks form is called *magma*. The best evidence we have indicates that it originates in chambers, or pockets, within solid rock at considerable depths below the surface of the earth. Estimates of the depth range from many thousands of feet to many miles. We do not know for certain how this rock material comes to be in a molten state. A likely explanation is that it has been melted as a result of heat produced by radioactivity (the spontaneous decay of radioactive elements in the course of which energy is given off in the form of heat). In any case, we know beyond a doubt that molten rock does exist beneath the surface from the evidence provided by volcanic activity, which we shall discuss in the next chapter.

Magma is a mixture of numerous components, some of which, at ordinary earth-surface conditions, exist as solids, some as liquids, and some as gases. In the first category are the minerals. In the last two categories water predominates, but also included are carbon monoxide, carbon dioxide, sulphur dioxide, and chlorine, as well as other elements and compounds. These are volatile substances which escape when the magma is solidifying. Our knowledge of their presence is based on analyses of the emanations from volcanoes. It is thought that as much as 10 per cent of some magmas comprises volatile material. These materials are important because many of them are chemically active and hence may react with preexisting rocks; they may carry minerals in solution; and they reduce the viscosity of the magma by their presence. Although igneous rock is formed by solidification of molten material, reversing the process by melting a sample of igneous rock to form a liquid does not make a magma, for the volatile materials would then be absent. The volatiles remain in the magma prior to crystallization because of the very great surrounding pressures and because there is ordinarily no exit by which they can escape to the surface through the miles-thick overlying layers of rock.

Origin of igneous rocks. Many aspects of the igneous process are still unknown to us. Igneous rock bodies differ markedly in the proportions of

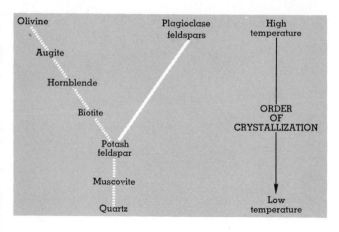

FIG. 7-10. Bowen reaction series.

their mineral constitutents. Some are granites, consisting largely of quartz and feldspar in approximately equal proportions; others are composed almost entirely of olivine or magnetite; and many are combinations of a relatively small number of minerals in various proportions. One possible, and perhaps most obvious, reason for the existence of igneous rocks of different composition is that they have formed from magmas of corresponding constitution. But substantial evidence suggests that this may not be the sole explanation. Detailed examination of igneous rocks, laboratory experiments, and theoretical studies of the chemistry of melts indicate that the events of igneous-rock formation occur in a regular, often-repeated sequence. As a magma begins to cool, certain minerals are the first to crystallize as solid grains. As time goes on and the magma cools further, other minerals solidify in a known sequence.

Two series of mineral crystallization documented by the American geologist N. L. Bowen are shown graphically in Fig. 7-10. Minerals that form first—i.e., at the highest temperatures—are at the top of the figure. As the temperature falls, the minerals shown lower on the chart crystallize. The melt that remains then reacts with the mineral grains already formed and changes their composition to that of the minerals still lower on the temperature scale. Note that the reaction series of the left branch, marked by dashed lines, is discontinuous: there are four distinct substances with none of intermediate composition. In contrast, the reaction series of the right branch is continuous, hence is marked by a solid line: the plagioclase feldspars are composed of calcium (Ca) and sodium (Na), in decreasing amounts of calcium and increasing amounts of sodium as we move from the top, or high temperatures, to the bottom, or low temperatures; thus they are not separable into individual substances of different composition. From the temperature scale on the right you will realize that the two series may proceed simultaneously—that is, at the same temperatures and hence at the same time, assuming that minerals of both series exist within a given magma. Whichever minerals do form, those grains that crystallize early will have crystal faces; and those that crystallize last, solidifying in the space between the already existing grains, will have the shape of the spaces rather than of their own crystal habit.

Some minerals that are of comparatively high specific gravity are among the first to form; being heaviest, they tend to settle to the bottom of the magma chamber. Materials of somewhat lower specific gravity may then crystallize and form a second layer. This may continue until all the magma has solidified. The process described is called *gravity separation*, shown in simplified form in Fig. 7-11. Examination of the igneous rock body thus formed would show the basal parts to contain a high proportion of early-crystallizing, high specific gravity minerals and the uppermost part to contain a high proportion of late-crystallizing minerals. A good example of igneous rock of this nature may be seen in the Palisades of New York and New Jersey.

How do the events described help to explain the variations in composition of the igneous rocks? Many geologists believe they provide an alternative or at least a complement to the possibility that igneous rocks of different composition have formed from differently constituted magmas. On the basis of Bowen's reaction series it is thought that groups of minerals which crystallize at the same time may be removed from the remaining magma by gravity separation or by other environmental stresses and thus form discrete rock bodies with little or no trace of the other substances with which they have coexisted in the molten form. Much more work remains to be done, however, before the nature of igneous-rock formation can be fully understood.

Classification of igneous rocks. The classification of igneous rocks is based on two characteristics—composition and texture, or grain size. Composition ranges from a preponderance of the dark-colored, high specific gravity minerals to all, or nearly all, light-colored, late-crystallizing minerals. A simple and workable method of classifying igneous rocks thus begins by dividing them into light colored and dark colored. The second criterion is grain size. If the melt cools rapidly, grains form simultaneously about many closely spaced centers and the resulting rock has many small grains (Fig. 7–12). If the magma is slow cooling, the slower rate of crystal growth allows time for molecules or ions from smaller crystals to migrate through the melt to larger crystals and thus build a small number of large grains (Fig. 7–13).

For purposes of texture classification igneous rocks are divided, somewhat arbitrarily, into *phaneritic* rocks, where the individual grains are large enough to be distinguished with the naked eye, and *aphanitic* rocks, where some kind of magnification is required. An additional designation

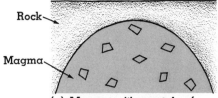

(a) Magma with crystals of the first mineral to separate

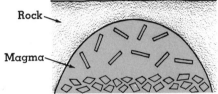

(b) Magma with first crystals settled to bottom as the second mineral crystals form

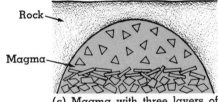

(c) Magma with three layers of crystals of different minerals

FIG. 7-11. Gravity separation. Successive stages, seen in cross section.

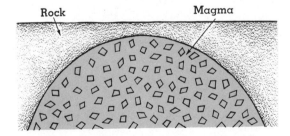

FIG. 7-12. Rapidly cooling magma forms many fine grains.

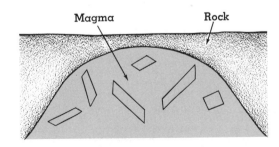

FIG. 7-13. Magma cooling slowly forms few large grains.

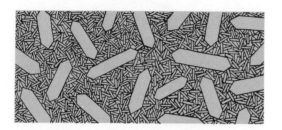

FIG. 7-14. Porphyry containing coarse grains and fine grains (full scale).

is needed because the rate of cooling of any rock body may not be uniform. A deeply buried magma, cooling slowly, will at first contain a comparatively few large grains in a melt. If for some reason the mixture of crystals and melt moves to a place where it cools more rapidly, the next grains to form will be more numerous and smaller. The resulting rock will contain some comparatively large grains surrounded by smaller grains. Rock composed of two or more distinctly different sizes of mineral grains is called *porphyry* (Figs. 7–14 and 7–15). Porphyritic rock may be entirely phaneritic or it may contain phaneritic grains in an aphanitic or glassy groundmass.

Molten material that has arrived at or near the surface as a result of volcanic activity may cool with such rapidity that it solidifies before molecules or ions in the melt can combine to make crystalline mineral grains. The result is formation of a natural glass. Obsidian is an example (Fig. 7–16).

Occasionally molten material may reach the surface before all the volatile substances have escaped. With the decrease in pressure they come out of solution and form a froth. If the material then solidifies too quickly to allow the gas bubbles to escape, cavities in the rock, called *vesicles*, are formed. Scoria and pumice are examples of vesicular rocks.

Table 7–3 shows a simple classification of igneous rocks. It is possible to construct a more complicated scheme showing rocks with gradations of

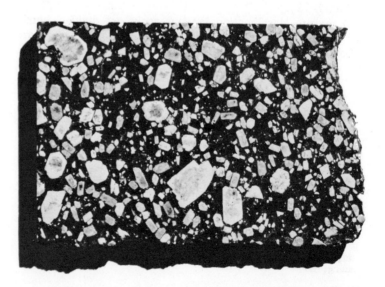

FIG. 7-15. Porphyry. Light-colored crystals showing faces in dark groundmass. (Ward's Natural Science Establishment, Inc.)

176

TABLE 7–3   CLASSIFICATION OF IGNEOUS ROCKS

TEXTURE	COMPOSITION		
	Light-colored minerals (feldspar, quartz dominant)		Dark-colored minerals (iron, magnesium dominant)
Phaneritic	Granite		Gabbro
Grains of two sizes		Porphyry	
Aphanitic	Felsite		Basalt (Scoria)[a]
Glassy	Obsidian (Pumice)[b]		

[a] Scoria contains large, roughly spherical vesicles.
[b] Pumice, usually light colored, contains small vesicles elongated during flow in many specimens.

FIG. 7-16. Obsidian. (Filer's.)

each property listed, but the classification then becomes somewhat unwieldy, and the simple outline shown is adequate for our purposes.

Because igneous rocks originate as completely continuous liquids (lacking pore space) under high pressure, they are, except for scoria and pumice, almost entirely solid material, with grains in close contact and interlocking. They are consequently strong, hard rocks of comparatively high specific gravity.

Characteristics of igneous rocks

1. Hard, strong, high specific gravity.
2. Isotropic; i.e., uniform in all directions and lacking oriented grains.
3. Irregularly shaped individual mineral grains everywhere in contact with adjacent grains.
4. Void spaces absent (except for scoria and pumice).

SEDIMENTARY ROCKS

Sedimentary rocks form as the result of the deposition of materials from fluids. They are deposited in many kinds of environments, and there is an extremely large variety of texture and composition within the group. Whether laid down by wind or by water, most sediments are deposited in layers of fairly uniform thickness, on comparatively level or gently

FIG. 7-17. Bedded and partly laminated sedimentary rock. Bed A: graded bedding (particles coarsen downward); bed B: nonlaminated; bed C: laminated; bed D: cross-bedded.

FIG. 7-18. Lamination in sandstones, Canyon de Chelly, Arizona. (National Park Service.)

sloping surfaces. The layers are the most conspicuous and most characteristic feature of sedimentary rock in large bodies (Fig. 7–17). *Laminations* are very thin layers produced by small variations in either grain size, mineral composition, or color (Fig. 7–18). They are by definition less than 1 centimeter in thickness and may be as thin as a sheet of paper. On a much larger scale, very thick alternating layers of different kinds of rock are also characteristic. These layers are called *beds*, or *strata*, and they may be many feet in thickness (Fig. 7–19). Laminations may or may not be present within a bed. Ordinarily, when laminations and beds do occur together they are planar and parallel to each other, although the spacing (or thicknesses) may be irregular. Some types of laminations are irregular in form and others are oblique to the beds.

Sedimentary rocks are grouped in three categories: *Clastic* sediments form as an accumulation of solid particles such as sand. *Chemical* sediments are for the most part the result of the evaporation of water that contains dissolved materials. *Organic* sediments are composed of tissues or skeletons of plants or animals more or less modified. Some life process is a requisite.

Clastic sedimentary rocks. Clastic sediments are composed of solid particles of preexisting rock or mineral grains that have been transported and deposited by a fluid—either air or water. They are classified primarily on the basis of the size of their particles. The terminology most generally in use in the Western Hemisphere was established by C. K. Wentworth, whose

178

FIG. 7-19. Bedding in sedimentary rocks. (National Park Service.)

FIG. 7-20. Conglomerate. Large, round varicolored pebbles cemented together in a lighter matrix. (Ward's Natural Science Establishment, Inc.)

FIG. 7-21. Breccia. Angular light-colored fragments in a dark matrix. (Ward's Natural Science Establishment, Inc.)

TABLE 7-4   WENTWORTH GRAIN SIZE CLASSIFICATION

PARTICLE NAME	SIZE RANGE (diameter in millimeters)	CLASTIC ROCK FORMED OF PARTICLES
Boulder	Over 256	Conglomerate
Cobble	256–64	
Pebble	64–4	
Granule	4–2	
Sand	2–1/16	Sandstone
Silt	1/16–1/256	Siltstone
Clay	Less than 1/256	Claystone and shale

system of classification is shown in Table 7–4. Note that there is no reference to mineralogic composition, size alone being the determinant. The individual particles in sandstone and conglomerate are large enough to be distinguished with the naked eye and are characteristically rounded to a greater or lesser extent (Fig. 7–20).

Along with this general scheme certain special names have come into use. Rocks formed of angular particles over 2 millimeters in diameter are known as *breccia* (Fig. 7–21). Rock composed of sand-sized particles of which over 25 per cent are feldspar is called *arkose*. *Claystone* and *shale*, which have the same particle size, are differentiated by the manner in which they break—claystone breaking into blocky fragments and shale into platy or linear pieces.

In many clastic sedimentary rocks, deposits of minerals have accumulated between the particles and act as a cement to form a hard, firm aggregate. Other clastics are held together only by compaction due to the pressure of surrounding material. The degree and material of cementation greatly affect rock strength, which may be exceedingly great for a sandstone well cemented by quartz or practically none for a shale held together by compaction only. Calcite is by far the most common cementing mineral; quartz is found with some frequency, and iron oxide occasionally. Some clastics have a clay *binder* in the interstices holding the particles together by adhesion. A binder differs from a cement in that its properties are affected by moisture: rocks with a clay binder are not as hard and firm when wet as when dry, whereas cemented rocks have approximately the same strength under both conditions. A sandstone that is not well cemented, or that has a clay binder from which sand grains rub off easily, is known as *friable* sandstone.

Chemical sedimentary rocks. Chemical sediments are formed by the precipitation of minerals from a water solution, most often because of evaporation. The crystalline grains thus formed either grow on the bottom of a body of water or settle to the bottom. Many chemical rocks are cemented by the same mineral that makes up the precipitated grains, hence are *monomineralic*. These are comparatively compact and dense. The following are some common chemical rocks, with their dominant mineral component: limestone (calcite), rock gypsum (gypsum), rock salt (halite), chert (chert).

Chemically deposited sediments may form in many ways, some not yet understood; consequently their classification is still a controversial subject. Further, some sediments may be boundary types—for example, limestone containing many fossils (organic material) in a chemical matrix.

Organic sedimentary rocks. The primary component of organic sedimentary rocks is some product of a life process, either animal or vegetable. Accumulation of the biologic material is the essential element in the origin of such rocks. Although there are a multitude of life forms, many environments of life, and a variety of ways in which the remains can be concentrated, the number of different organic sediments is surprisingly small. Several fuels are formed from plant life: *peat*—slightly compressed, partially altered preserved vegetable matter; *lignite*—brown coal; *bituminous coal*—soft coal; and *anthracite*—hard coal. In the order listed, these fuels come closer and closer to being pure carbon and contain correspondingly less of the volatile constituents. Diatoms—microscopic simple plants—secrete skeletons of silica or opal, which accumulate as very fine sediments on the bottoms of bodies of water to form diatomaceous earth, or *diatomite*. Some other plants secrete calcite in their tissues or on external surfaces to form *organic limestone* (also formed by animal secretions, as in coral reefs). Most such rocks have been formed by primitive plants called algae. Animal skeletons are the primary constituents of many organic sediments. Microscopic organisms contribute their skeletons in such great numbers that they may constitute over 90 per cent of a sediment. Many hundreds of feet of rocks may be composed of skeletons of Foraminifera (calcite) or Radiolaria (opal).

Importance of sedimentary rocks. Because of the many different origins and combinations of materials of the sedimentary rocks, their categories are not precise and may even overlap. Also, their diversity makes it difficult to identify features common to the entire group, as you can see from the list below. But one significant fact that is true for sedimentary rocks as a whole gives them a unique importance to geologists as well as to all those who are curious about the world around them. *Sedimentary*

FIG. 7-22. Gneiss, a metamorphic rock showing parallelism of grains, which are perpendicular to the direction of the force applied. (Ward's Natural Science Establishment, Inc.)

rocks are formed entirely at the surface of the earth. Thus they give us evidence of past events at the surface—evidence that is accessible to direct observation and from which we can construct a chronology of change. In a later chapter we shall discuss earth history and examine some of the methods by which geologists read the record of the past in the strata of the sedimentary rocks.

Characteristics of sedimentary rocks

1. Lamination and/or bedding conspicuous in some.
2. Fossils of plants or animals present in some.
3. Many are soft, weak, and comparatively low in specific gravity.
4. Coarser clastic grains are usually rounded, may have voids between them.

METAMORPHIC ROCKS

Metamorphic rocks are preexisting igneous or sedimentary rocks that have been altered by environmental change to a form distinctly different from that of the original rock. Metamorphic transformation may be produced by changes in temperature, pressure, or chemical action—or any combination of these—of an intensity and character generally occurring beneath the surface of the earth. The essential phases of the metamorphic process may be shown schematically as follows:

change in $\longrightarrow$ rock unstable in $\longrightarrow$ change $\longrightarrow$ rock in equilibrium
environment   new environment   in rock   in new environment

The process begins with an increase in heat or pressure, or a change in the surrounding chemical balance, such as to create instability in the chemical and physical structure of the rock. Note, however, that this result is a function of the nature of the rock as well as of the environmental change. For example, granite is formed initially under conditions of high temperature and high pressure, and it will be stable thereafter in all similar environments regardless of any intervening environmental changes; limestone, on the other hand, is initially formed under low-pressure–low-temperature conditions at the earth's surface, hence will be substantially modified in any high-temperature–high-pressure environment. The end product of metamorphism is always a rock whose chemical and physical structure is relatively stable in the changed environment. Metamorphic rocks that we see at the surface are more or less as they were when formed in the depths because they react very little and very slowly to the moderate conditions obtaining at the surface.

The heat that is involved in metamorphism may derive from a nearby igneous body (even close to the surface), or it may come from the interior of the earth. Pressure may originate from either of two possible sources—the weight of a thick layer of rock bearing down on a deeply buried material; or a horizontal force of unknown origin accompanying deformation of the earth. Heat and high pressure are often associated, and we may therefore consider their effects when acting in combination. These agents together characteristically produce an increase in the size of mineral grains; rotations and reorientation of mineral grains (Fig. 7–22); and sometimes mineralogic changes, including loss of some substances. The process of change is gradual, and the products often grade into one another. Foliated metamorphic rocks include *slate*, a clay-size rock which cleaves into flat plates because of parallelism of platy grains; *schist*, a rock containing a large percentage of minerals with oriented coarse grains of cleavable minerals; and *gneiss*, a rock containing alternate bands of schistose and noncleavable mineral grains.

Under conditions in which heat or chemical action predominate, the resulting rocks lack oriented grains and sometimes have substantially altered mineral content. Where heat acts alone, for example, limestone becomes *marble*, also composed of calcite but with coarser grains and often in purified form. Chemical action may introduce new compounds and thus either add minerals or replace some with others. Where the predominant mineral is quartz, the resulting rock is called *quartzite*. Some confusion may result because this name is employed both for metamorphic rocks and for the sedimentary rocks formed by cementation of a quartz sand with a quartz cement. The origins of these two kinds of quartzite may be recognizable and should be stated as part of the rock identification. In some instances, however, the origin may not be apparent. In any case, a specimen should be identified as quartzite if it is virtually all quartz and if when it is broken the fracture surface passes through the individual quartz grains rather than around them.

Because metamorphic rocks form under comparatively high temperature and high pressure, they resemble igneous rocks in that most are hard, strong, and of comparatively high specific gravity.

Characteristics of metamorphic rocks

1. May be banded; platy or elongate grains parallel to bands.
2. Usually hard, strong, with high specific gravity.
3. Void spaces absent.
4. May show features of igneous or sedimentary rocks in various stages or degrees of alteration or destruction.

THE ROCK CYCLE

Although we have discussed the three major rock groups individually, for the sake of clarity, you will no doubt have realized that there are im-

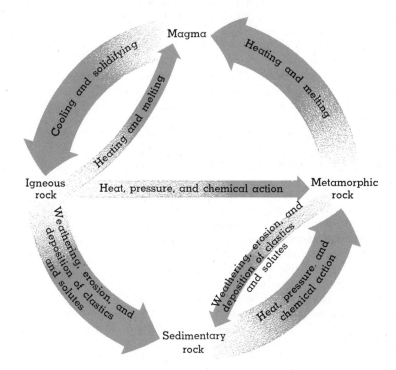

Magma

Cooling and solidifying

Heating and melting

Heating and melting

Igneous rock

Heat, pressure, and chemical action

Metamorphic rock

Weathering, erosion, and deposition of clastics and solutes

Weathering, erosion, and deposition of clastics and solutes

Heat, pressure, and chemical action

Sedimentary rock

FIG. 7-23. The rock cycle, greatly simplified. Remember that magma is not just melted rock but contains volatiles as well. Weathering produces clastics and dissolved material as a step prior to compaction of a sedimentary rock. Sedimentary rock can be weathered and the solid and dissolved products transported and deposited to make other sedimentary rock.

portant interconnections among them. We may express these connections in their simplest form as follows:

1. Sedimentary rocks are formed from materials of other types of rock.
2. Metamorphic rocks are formed from other types of rock that are vulnerable to certain environmental conditions.
3. Igneous rocks are formed from magma that may contain molten material from other types of rocks.

Figure 7–23 shows graphically the sequence of events in the rock cycle and the changes they produce. The cycle may occur either in its complete form or with parts by-passed, as indicated by the arrows. Remember that no rocks we presently observe are considered by geologists to represent the original rocks of the earth because over the ages since the planet was formed many such cycles of change have taken place.

SUGGESTED REFERENCES

Fenton, C. L., and M. A. Fenton: *The Rock Book*, Doubleday & Company, Inc., Garden City, N.Y., 1940.

Holden, Allen, and Phyllis Singer: *Crystals and Crystal Growing*, Doubleday & Company, Inc., Garden City, N.Y., 1960. (Paperback.)

Hurlbut, Cornelius S., Jr., and H. E. Wenden: *Changing Science of Mineralogy*, D. C. Heath & Company, Boston, 1964. (Paperback.)

Leet, D. J., and Sheldon Judson: *Physical Geology*, 3rd ed., Prentice-Hall, Inc., Englewood Cliffs, N.J., 1965.

Pearl, Richard M.: *Rocks and Minerals*, Barnes & Noble, Inc., New York, 1956. (Paperback.)

Pettijohn, F. J.: *Sedimentary Rocks*, 2nd ed., Harper & Row, Publishers, Inc., New York, 1957.

Rogers, John J. W., and John A. S. Adams: *Fundamentals of Geology*, Harper & Row, Publishers, Inc., New York, 1966.

Tyrell, G. W.: *The Principles of Petrology*, E. P. Dutton & Co., Inc., New York, 1929.

VULCANISM, TECTONISM, EARTHQUAKES AND THE EARTH'S INTERIOR

Igneous rocks are formed by solidification from magma, the molten rock material that exists at depths beneath the surface of the earth. But we have said that the igneous rocks may solidify either in the depths or at the surface. How is this possible? Under certain conditions magma may be expelled from its original location in the depths, traveling up through the surrounding rock and erupting at the surface, often under great pressure. There, at the relatively low surface temperatures, it quickly solidifies to form a variety of igneous rock types.

Before describing the products of this process, let us clarify some important terms and relationships:

Vulcanism is the formation of all the various kinds of igneous rock, at whatever location.

Intrusive igneous rocks are those that are formed in the depths.

Extrusive igneous rocks are those that are formed at the surface.

Lava is magma which reaches the surface and erupts.

Volcanic activity is the emission of lava and the building up of volcanic structures thereby.

Intrusive *igneous process*	*Extrusive* *igneous process*
magma ⟶ igneous rock	(magma) lava ⟶ igneous rock

EXTRUSIVE IGNEOUS ROCKS

The products of the extrusive igneous process are of special interest to us for their importance in shaping the face of the earth. Intrusive igneous rock is visible only to the extent that it is uncovered by the erosion of overlying materials. But the extrusives are entirely surface features, serving to build up landforms in what we call an *aggradational* process. Thus a volcanic peak or a lava flow is an aggradational igneous landform. There are several kinds of volcanic eruptions, and each constructs a volcanic peak that is of characteristic form and composed of characteristic materials. We may make the general statement, however, that since lava always cools and solidifies comparatively rapidly under the impact of surface conditions, the extrusive rocks are all of aphanitic or glassy texture, except for a small number that are porphyritic with an aphanitic matrix.

Shield volcano. Lava of low viscosity will often flow out at the surface in an unexplosive manner and spread out in a relatively thin sheet which, when solidified, makes a broadly conical deposit. Repeated extrusions of the same nature enlarge the cone, but it retains a gently sloping surface,

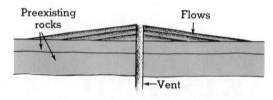

FIG. 8-1. Cross section of a shield volcano.

usually 4 to 10° from the horizontal. Thus a quiet eruption produces the *shield volcano*, so called because of its gentle slopes and broad extent (Fig. 8–1). This type of volcano is usually of basaltic composition. The Hawaiian Islands are composed largely of shield volcanoes, the highest of which is Mauna Loa.

Cinder cone. When the magmatic material that reaches the surface is of high viscosity, it is likely to plug the vent upon cooling. If the internal pressure builds up sufficiently, the plug is expelled in a violent manner, and a jet or spray of hot fluid erupts explosively. Much of the ejected lava solidifies while still in the air and falls back to the ground as solid particles, known as *pyroclastics* (from the Greek: *pyro*, fire; *clastic*, broken) (Fig. 8–2). These particles are found in a variety of sizes and shapes, which we classify as follows: (1) small particles of irregular shape are called *ash*, less than 4 millimeters in diameter, and *lapilli*, 4 to 32 millimeters; (2) larger particles are *blocks*, of angular form, and *bombs*, spherical or elongate, with surfaces smoothed by air resistance (Fig. 8–3). The pyroclastics accumulating around the vent build up deposits in layers at the angle of repose—about 34° from the horizontal—and thus form the steep-sided *cinder cone* (Figs. 8–4 and 8–5).

Composite cone. Although many volcanoes extrude material of uniform composition throughout their active life, there are some in which the lava is of alternating or gradually changing composition. The resulting landform is thus made up of a mixture of materials, and is likely to be of intermediate slope, between the low angle of the shield volcano and the high angle of the cinder cone. In many of these *composite cones* the slopes are concave upward, which suggests that the later eruptions were explosive.

It is interesting to note that one of the most beautiful mountains in the world, Fujiyama, in Japan, is a volcano of the composite type (Fig. 8–6). The grace and symmetry of its slopes have so captured the imagination of the Japanese people that it appears as a symbol throughout Japanese poetry, legend, and art.

Plateau basalts. In some areas of the world, igneous extrusions have formed lava flows rather than mountains. These are extensive deposits, essentially horizontal, occurring one on top of another for a total of hundreds or even thousands of flows that range in thickness from a few feet to several hundred feet each. Pyroclastics and sedimentary rocks may be interbedded with the lava flows. The flows appear to have come to the surface as quiet extrusions from numerous vents or fissures, and the interval between flows may apparently be of long or short duration. All are of basaltic composition, hence they are called *plateau basalts*. The thicknesses and total volumes are immense. The Deccan flows of India average

FIG. 8-2. Building partially covered by pyroclastics, Kilauea Iki eruption, Hawaii, 1959. Shelter is more than one-half mile downwind from source. Steps formerly led up to shelter from ground level. (National Park Service.)

FIG. 8-3. Volcanic bomb. Note elongation. (Filer's.)

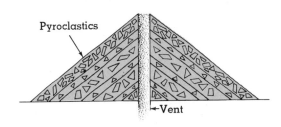

Pyroclastics

←Vent

FIG. 8-4. Cross section of a cinder cone.

FIG. 8-5. Oblique air photo of two cinder cones with craters (light-colored slopes), Peru. One cone, near upper-left corner, breached by lava. Flows moving toward observer and between cinder cones. (Aerial Explorations, Inc.)

2000 feet in thickness and total 75,000 cubic miles in volume. In the Columbia Plateau of Washington, Oregon, and Idaho, the flows reach a maximum of 5000 feet in thickness (including some sediments) and have a total volume of about 20,000 cubic miles.

Craters and calderas. In most volcanoes the vents are enlarged at the top. The hollow thus formed at the summit of the peak is called a *crater*.

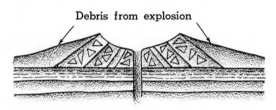

FIG. 8-7. Caldera of explosive origin, cross section.

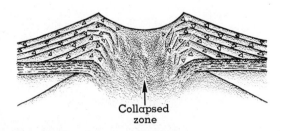

FIG. 8-8. Caldera of collapse origin, cross section.

Where the enlargement is so great that its width is much greater than the depth of the vent (at least three times as great, according to some authorities), we call the resulting basin-shaped depression a *caldera*. There are calderas that are so very much larger than most craters that they are thought to be of different origin. Some no doubt were created by violent explosions; in many cases this can be proved by demonstrating that the volume of debris piled around the caldera equals the volume of material missing from the depression (Fig. 8–7). In other instances there is not sufficient fragmental material to support the hypothesis of an explosion. Here the most likely cause is the collapse of the volcano and its foundation following the withdrawal of the magma (Fig. 8–8).

Types of lava. Lavas of low viscosity, which are called *pahoehoe*, solidify with smooth surfaces. Lava that is more viscous, either because it is of different composition or because it has cooled and in part solidified, is known as *aa*; it moves in the form of solid blocks in a thick liquid, and will form a flow having an extremely rough surface. The names are of Hawaiian origin.

C. E. Dutton, an American geologist who was the head of the division of volcanic geology of the U.S. Geologic Survey at the end of the nineteenth century, was responsible for these graphic descriptions of the two types of lava: "The first form is called by the Hawaiians pa-hó-e-hó-e. Imagine an army of giants bringing to a common dumping-ground enormous cauldrons of pitch and turning them upside down, allowing the pitch to run out, some running together, some being poured over preceding discharges, and the whole being finally left to solidify. The surface of the entire accumulation would be embossed and rolling, but each mass by itself would be slightly wrinkled, yet, on the whole, smooth, involving no further impediment to progress over it than the labor of going up and down the smooth-surfaced hummocks. The second form is called a-a, and its contrast with pahoehoe is about the greatest imaginable. It consists mainly of clinkers sometimes detached, sometimes partially agglutinated together with a bristling array of sharp, jagged, angular fragments of a compact character projecting up through them. The aspect of one of these aa streams is repellent to the last degree, and may without exaggeration be termed horrible. For one who has never seen it, it is difficult to conceive such superlative roughness."

Flows formed of aa lava may occur either as extensive sheets or as fingerlike "rivers." Such flows may cool and solidify on their surface, which is in contact with the air, while the lava in the interior is still fluid. If the hardened surface layer is subsequently broken and the unsolidified material flows out, *lava tunnels* are formed, extending for as much as several hundred feet.

The intrusive igneous rocks are formed from magma that cools and crystallizes beneath the surface, either within its original chamber or upon injection into openings in the surrounding rock. Insulation provided by overlying rock allows for a slow cooling process and the formation of large grains. For this reason few intrusive rocks are aphanitic or glassy in texture.

Because they form in the depths, the intrusive rocks remain unseen until they have been uncovered by the processes of erosion or by deep excavation. The great thickness of cover that must be removed before they can be observed requires great uplift, so most large exposures of intrusive igneous rock occur in the central parts of mountain ranges or in other areas where there has been uplift of many thousands of feet.

Batholiths and laccoliths. When magma cools, crystallizes, and hardens in its original chamber, the resulting rock body is likely to have a rounded upper surface. Thus a very small section will at first be visible as erosion removes the overlying material; then, as the land surface is brought lower and lower, more of the igneous rock will be exposed. We may observe it at any point after the initial exposure of the small upper section. If we see such an intrusion and the evidence suggests that the rock body extends downward indefinitely, we term it a *batholith* if its exposed area is greater than 40 square miles and a *stock* if it is less. Most such intrusions are composed of granite.

A batholith may, during intrusion, arch the overlying rocks upward without cutting across them; or it may enlarge the chamber by cutting through surrounding rock, causing fragments to break off and sink into the magma. These are known as *concordant* and *discordant* batholiths, respectively (Fig. 8–9). In the case of a discordant batholith the breaking off of sections of surrounding rock is called *stoping*, and the fragments, which become part of the intrusive rock body, are known as *xenoliths* (Fig. 8–9).

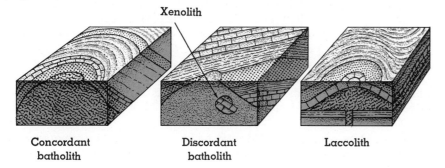

Xenolith

Concordant batholith   Discordant batholith   Laccolith

FIG. 8-9. Block diagrams of intrusives.

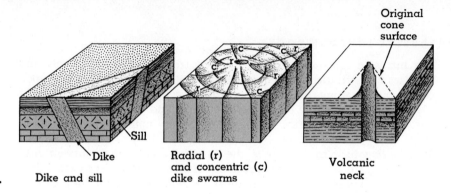

Original cone surface

Sill

Dike

Dike and sill

Radial (r) and concentric (c) dike swarms

Volcanic neck

FIG. 8-10. Block diagrams of igneous structures.

When magma leaves its chamber and moves upward through a conduit, it may reach a stratum of overlying rock that is impenetrable, push it upward, and fill the lens-shaped space that its pressure has created. The resulting rock body is called a *laccolith* (Fig. 8–9). If you will imagine observing a batholith and a laccolith from above, you will realize that they may be very similar in appearance. We must look for evidence that the rock body has a lower limit, with rock of a different character below it, in order to determine that it is a laccolith. Most laccoliths that have been observed are much smaller than batholiths, few being more than a few miles in diameter.

Other igneous structures. Magma may be injected between beds of sedimentary rock, where it cools and solidifies to form sheetlike rock bodies parallel to the beds. These are called *sills* (Fig. 8–10). Where an intrusion cuts *across* sedimentary beds, metamorphic foliation, or another igneous rock body, the structure is known as a *dike* (Figs. 8–10 and 8–11).

Dikes are often found in groups, or *swarms*, for where there is one extensive planar zone of weakness in the rock, others are likely to exist; and if magma enters one such zone, it is likely to enter the others. Dike swarms may be parallel, radiating from a center, or concentric about a center (Fig. 8–10). Since intrusions may take an irregular course, a sill may change direction to become a dike, and the two will form a continuous igneous rock body.

After a volcano has become inactive, lava may solidify in the vent. Later erosion may remove the least resistant materials, leaving the igneous filling of the vent standing high in the air. This is called a *volcanic neck*, or *volcanic plug* (Figs. 8–10 and 8–11).

The problem of granite. Granitic rocks, which contain minerals characteristic of the low-temperature end of the Bowen reaction series (for example, quartz, muscovite, and potash feldspar), are exposed in the cores of many present and past continental mountain ranges. Basalts, which are characterized by minerals crystallizing at the high-temperature end of the Bowen reaction series, are found associated with the ocean basins and as extensive plateau eruptives on the continents. Igneous rocks of intermediate composition are comparatively rare. This is a puzzling situation, because if we assume that the various igneous rocks form from magmas of different composition, we would expect, by normal statistical distribution, that the intermediate rocks would be the most abundant, and the extremes (granite and basalt) the rarest; yet, it is actually the granitic and basaltic rocks that predominate. Geographic and compositional distributions suggest that magma composition is not a statistical phenomenon, that is, that intermediate composition is not typical. How then is the abundance of basalt and granite to be explained?

One explanation is based on the process of gravity separation, discussed

FIG. 8-11. Shiprock, a volcanic neck and associated dike in northeastern New Mexico. (The American Museum of Natural History, New York.)

in Chapter 7. As a result of gravity separation, a batholith might grade from a granite in the uppermost part, through intermediate rocks in the middle, to a gabbro at great depth. But when a basaltic magma is analyzed it is found that only 5 to 10 per cent of it can go into the making of a granite. Therefore one possible explanation of the great area of granite and granitic rocks exposed on the continents would be that in nearly every batholith only the uppermost part has been exposed by erosion, and that nowhere has erosion removed the comparatively thin upper layer of granitic rock and exposed the intermediate igneous rock. The lack of extensive outcroppings of igneous rock of composition between granite and gabbro supports the conclusion that intermediate magmas are not the most abundant.

An alternative hypothesis has been suggested to the effect that granite forms by basaltic magma intrusion of other rock and melting of the intruded rock to form a mixture having the composition of granite.

Substantial evidence is found to indicate that granite can originate by metamorphism, without having had a molten stage at any time in its history. Such a process is called *granitization*. Many sedimentary rocks are rich in minerals characteristic of granite or have compositions chemically close to that of granite. If put in an environment of high pressure and high temperature, these may be altered, with or without going through the liquid state, to a rock which can be called a granite because of its composition and texture. Some authorities believe that as much as 85 per cent of all granite is formed in this way. It is unquestionable, however, that some granites form in other ways than by granitization, for some show evidence of having been injected as a fluid—evidence such as disturbance of the preexisting rock, cross-cutting relations with it, and injection of fluid which solidifies in the form of dikes and sills. Some granites do not exhibit any clear marks of their origin. So the problem is as yet largely unsolved.

TECTONISM AND GEOLOGIC STRUCTURE

We generally think of the earth as solid, strong, and stable, but it can at times be most active and mobile. When the earth "quakes," there is a rapid and violent release of energy and very substantial motion. After the 1964 earthquake near Anchorage, Alaska, for example, it was found that a 30,000-square-mile area had sunk up to 6 feet and an equal area had been uplifted, at one point to a height of 33 feet. In contrast, very gradual changes of altitude, accomplished over hundreds of thousands, or even millions, of years, are shown by the existence in mountaintop sedimentary rocks of fossils of animal and plant species whose habitat is the ocean bottom. In addition to these expressions of the earth's vertical mobility, we also see that surface and near-surface materials can move horizontally. Some streams in California, for example, are known to have been offset as much as 450 feet. We cannot, then, consider the earth to be a rigid, static sphere of rock. How can its movements and deformation be explained?

All solid materials have elastic properties—that is, they may be altered by a deforming force but once the force is removed they will return to their original shape. Some solids, however, may undergo permanent deformation. Putty and modeling clay when changed by an external force will retain their altered shape after the deforming force has been removed. We say they are *plastic* materials. Plasticity is a property of some kinds of rock under some conditions. The mineral calcite reacts to a force in a manner somewhat analogous to putty; thus when deformed slowly limestone may show plastic deformation. For a good example, consider the limestone benches installed in English burial plots several centuries ago. The seat slabs, supported by legs at or near the ends, have developed a sag in the center under the force of gravity. Of course, if the benches were turned upside down and left for another few centuries, the seats would gradually return to their original shape. But only a force in the opposite direction would reverse an otherwise permanent change in shape.

How does calcite deform permanently? Let us look at a rough model of crystalline molecular structure, greatly enlarged, to answer this question. Figure 8–12 shows a block of crystalline material. One corner is firmly against a buttress B. A force is applied at F, as indicated by the arrow. The rows of molecules glide, row past row, distances which are multiples of the spacing between adjacent molecules as shown. This produces permanent deformation without rupture, without altering the molecular arrangement, and without forming a different mineral. We believe that all crystalline materials can potentially be deformed in this way. Laboratory experiments have shown that it is possible under certain conditions to deform plastically many minerals that appear to be elastic.

In plastic deformation, a material is changed in shape but retains its unity. Brittle deformation involves a rupture, in whole or in part. Glass, cast iron, ice, and many other solids will break instantly into many fragments if struck a sufficiently powerful blow. Limestone (composed of calcite) can undergo brittle failure. How can calcite have the ability to deform both plastically and as a brittle material? If you think of our examples you will realize that the decisive factors are the duration and magnitude of the force applied. Many crystalline materials deform as brittle matter under a great force rapidly applied and as plastic matter under a moderate force applied over a long period. So we find that some kinds of rock have deformed as brittle materials under large forces rapidly applied

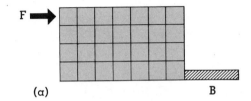

(a)

B

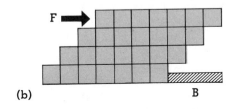

(b)

B

FIG. 8-12. Deformation of crystalline materials by gliding.

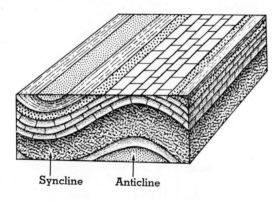

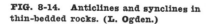

Syncline   Anticline

FIG. 8-13. Block diagram showing major fold types in sedimentary rocks.

near the surface of the earth and have also deformed as plastic material under smaller forces applied more slowly far below the surface.

The strength of a particular rock structure—its ability to withstand deformation of any kind—is a function of the scale of the structure as well as of the material of which it is composed. One thinks of granite as a strong material, and when used in the construction of buildings, bridges, and other relatively small structures, it is. But the rocks in the earth exist in much larger bodies. A pillar of granite with a crushing strength of 20,000 pounds per square inch can be about 20,000 feet tall without crumbling at the bottom from its own weight; but if it is much taller than that, the bottom part will fail under the weight of the upper sections. Rock structures exist in the earth in thicknesses much greater than 20,000 feet, and the force of gravity causes the upper sections to press on those below, creating what we call *confining pressure*. Halite is quite plastic under such pressure; below a depth of only 4000 feet it will flow under the weight of the overlying rock. Limestone (calcite) is somewhat stronger, granite strongest of all; but all will fail under sufficiently rigorous conditions.

STRUCTURES PRODUCED BY PLASTIC DEFORMATION

When bedded sedimentary rock fails plastically as a result of environmental stress, a fold forms in the rock material. Consider a sheet of paper held horizontally and flexed upward in the middle—and another sheet of paper flexed downward in the middle. An upward fold in rock material is known as an anticline, a downward fold as a syncline (Figs. 8–13 and 8–14).

Folds occur in a wide range of sizes. Some that have been measured in metamorphic rock are only a few inches in each direction (Fig.

FIG. 8-14. Anticlines and synclines in thin-bedded rocks. (L. Ogden.)

8–15), but there are giant anticlines and synclines several hundred miles long and involving vertical motion of 20,000 feet. Further, imagine the wrinkles in a tablecloth thrown on a polished tabletop—they represent very well one property of folds; they do not go on indefinitely but gradually diminish in both directions and eventually die out altogether.

The two sections of a fold on either side of the axial plane (the plane of sharpest folding) are called *flanks*, or *limbs*. Folds may be symmetrical—where one flank is the mirror image of the other—or asymmetrical. Several kinds of folds are shown in Fig. 8–16. The inclination of the flanks may be so slight as to be imperceptible without the use of surveying instruments or so large that the tilt is to the vertical position or even beyond; in the latter case we speak of an *overturned* fold. Folds are shown in an air view in Fig. 8–17.

What kinds of deforming forces cause rocks to fold? They may be purely vertical forces, as where the intrusion of a batholith or laccolith produces pressure upward and creates an anticline. Or they may be opposing horizontal forces, which flex the strata and form anticlines and synclines. The opposing horizontal forces may be of unequal magnitude, and the resulting folds will then be asymmetrical or overturned. Folds may be formed even where the horizontal forces are not in direct opposition to each other. Two such forces are called a *couple*. Their action is shown in Fig. 8–18. In (a), two forces applied at A and B are shown as they affect a rectangular area. In (b), the rectangle has been deformed to a parallelogram. Notice how the diagonal A-B has been shortened. Folds elongated in the C to D direction will form because of compression between A and B.

It is essential to be able to map the structure of deformed rocks. Geologic structures can be represented on a map just as man-made

FIG. 8-15. Folding in a gneiss. Anticlines and synclines only a few inches in maximum dimension. (The American Museum of Natural History, New York.)

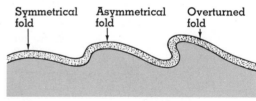

Symmetrical fold   Asymmetrical fold   Overturned fold

FIG. 8-16. Fold types in cross section.

FIG. 8-17. Concentric bands representing sedimentary rock units exposed on ground surface and viewed from the air. Radar image of Tuskahoma Syncline, Oklahoma. Shading as if illuminated from top of image. (National Aeronautics and Space Administration.)

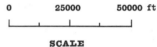

0   25000   50000 ft

SCALE

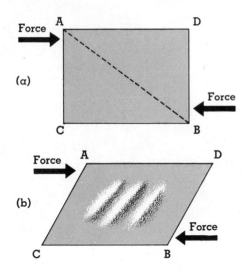

(a)

(b)

FIG. 8-18.
Map view of folds formed by a couple.

structures can, but they require the use of special concepts. In mapping folds (as well as other planar features) we use the terms *dip* and *strike*. The dip refers to the inclination of the tilted bed, and contains two components: the angle of inclination and the direction of the inclination. We measure the angle down to the inclined plane starting from the horizontal plane, as shown in Fig. 8–19. The greatest possible dip is 90° (the vertical position). To describe the dip completely, i.e., giving angle plus direction, we state, "a dip of 40° to the northwest," or "a dip of 85° to the south." The strike is the line of intersection of the inclined plane with the horizontal plane; it is the compass direction of this line that we record. In Fig. 8–20, which includes a north arrow for reference, we see that the direction of the intersection of the horizontal plane with the inclined bed is approximately south-southwest, and the strike would be given as "S 22° W." The complete description of such an inclined bed at a point might then read

strike S 22° W, dip 18° SE.

It would be shown on a map by the symbol you see in Fig. 8–21. The crossbar of the T is drawn in the exact direction of the strike, the staff of the T points in the direction of the dip, and the number is the angle of the dip. The symbol is so placed on the map that the intersection of the staff and crossbar falls exactly at the point where the measurement was made.

STRUCTURES PRODUCED BY BRITTLE DEFORMATION

The brittle failure of rock involves its rupture, sometimes accompanied by motion of the fractured rock sections and sometimes not. A fracture without subsequent movement parallel to the fracture surface is known

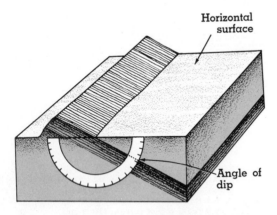

FIG. 8-19. Dip of a bed of rock in block diagram.

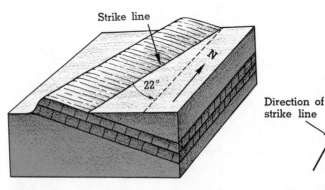

FIG. 8-20. Dip and strike of a bed in block diagram.

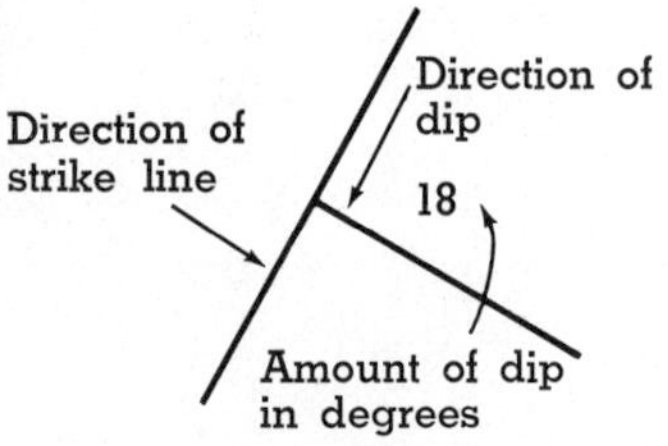

FIG. 8-21. Symbol used to indicate dip and strike on a map.

as a *joint*. As an illustration, consider the parallel cracks that would appear if a sheet of safety glass were bent in the shape of an anticline. The cracks would be parallel to the axis of the fold and would be exactly analogous to joints. Joints are common in bedded sedimentary rock near and roughly parallel to the axes of anticlines and synclines, for it is in these locations that the rock material has been folded most intensely (Fig. 8–22). Igneous rock may shrink upon solidifying and so become jointed (Fig. 8–23). In basalt flows, conspicuous joints in a more or less regular honeycomb pattern may occur, forming prisms perpendicular to the long dimensions of the flows (Fig. 8–24). The spectacular prisms of the Giant's Causeway in Ireland are examples of prismatic jointing. Since joints are planar features, they are described in terms of dip and strike.

Where a rock has many joints parallel to each other, we speak of a *joint set* (Fig. 8–25). In massive, homogeneous rock units such as batholiths, joints often occur in three sets at approximately right angles to each other; but sets may also be oblique to other sets. In bedded rocks one or more joint sets may be present, with most sets perpendicular to the bedding.

A *fault* is a rupture with further motion; the fracture is followed by slipping along the break to produce offset of the rock units. The displacement may be so slight as to be barely perceptible or it may be many miles. The break may result in a very smooth plane or an undulating surface; but so few faults are true planes that it is best to speak of fault surfaces rather than planes. Nevertheless, a fault surface has dip and strike—but they are not the same at all points. Because of the friction associated with the movement of the rock sections, the rock surface at a fault is likely to be polished and is often grooved. A polished surface is called *slickensided* (from the English dialect word for smooth).

Some of the features of a fault and the terms we use to identify them are illustrated in Fig. 8–26. Bedded sedimentary rocks are shown before rupture and movement and as they appear after faulting. Arrows indicate the relative motion of the two blocks. A tunnel is shown along the fault surface. A person in the tunnel would be standing on rock beneath the fault (the right-hand block), which for this reason is called the *footwall block*. Correspondingly, rock overhead in the block above the fault (the left-hand block) is called the *hanging wall block*. We speak of a *normal* fault if the hanging wall block moves downward with respect to the footwall block, the movement having occurred as a result of tensions within the rock (Figs. 8–26 and 8–27). If, on the other hand, horizontal compression is the moving force, the motion will be the reverse (Fig. 8–28). Hanging wall and footwall blocks will remain in the same relative positions, but the hanging wall block will move up with respect to the footwall block and ride over it. In this case we speak of a *reverse* fault. If the fault surface

FIG. 8-22. Joints concentrated near crests and troughs of folds, cross section.

FIG. 8-23. Prismatic joints in an igneous body. Devil's Tower, Wyoming. (The American Museum of Natural History, New York.)

◀ FIG. 8-24. Prismatic jointing in a lava
flow. Devil's Postpile, California.
(National Park Service.)

▲
FIG. 8-25. Joints in sandstone,
Canyonlands National Monument.
(National Park Service.)

VULCANISM, TECTONISM, EARTHQUAKES

203

Before rupture

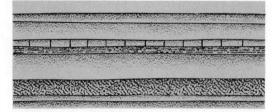

After rupture and motion

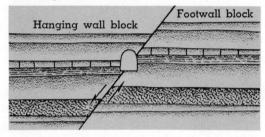

Hanging wall block

Footwall block

FIG. 8-26. Normal fault in cross section.

FIG. 8-27. Normal fault. Block on left downdropped exposing fault surface. (U.S. Coast and Geodetic Survey.)

has a low angle of dip (less than 20°), it is called an *overthrust*. A fault may not involve any vertical movement. If one block moves horizontally, *slipping* along beside the other parallel to the *strike* of the fault, we have a *strike-slip* fault (Figs. 8–29 and 8–30). It is relatively rare to see a fault that involves movement in the vertical or horizontal direction alone.

Although a fault may be unaccompanied by other faults within the same rock body, it is more usual to find that the forces at work have created more than one rupture. Internal tensional forces may cause (1) a *graben:* two faults enclosing a block downdropped with respect to the other blocks on either side (Fig. 8–31); or (2) a *horst:* two faults with the block between upthrown with respect to the blocks on either side (Fig. 8–32). Figures 8–31 and 8–32 show faults where the movement has been so rapid that the fault makes a distinct scarp, or cliff, but this is very often not the case. Faults may be associated with folds, or with other faults. Reverse faults and overthrusts are often found together with over-

After rupture and motion

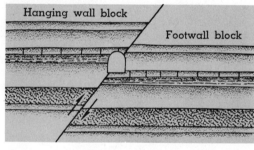

Hanging wall block

Footwall block

FIG. 8-28. Reverse fault in cross section.

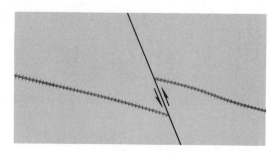

FIG. 8-29. Strike-slip fault. Map view showing railroad offset by fault. Arrows indicate relative motion.

turned folds, or we may see a fault that changes gradually into a fold along its strike (Fig. 8–33).

Thus far we have been considering faults characterized by a single fault surface, but we may also see a *fault zone* many feet thick, in which the rock is badly crushed. The broken rock material may then provide a conduit for the relatively easy passage of fluids; and if the passing fluids come from a magma and carry dissolved minerals, the fault may become the site of an ore deposit. But where the rock is very finely crushed, referred to as a *gouge*, it may be altered to a clay, and it then becomes a barrier rather than a conduit for fluid movement. Faults may occur subsequent to the deposition of ores, and many profitable mines have had to be abandoned when the ore was found to be cut off by a fault.

A word of caution about the geologic structures we have been discussing: Remember that joints, folds, and faults extend downward into the bedrock below the surface—or perhaps we might say, up through the bedrock.

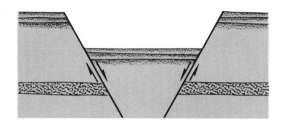

FIG. 8-31. Graben in cross section.

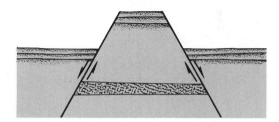

FIG. 8-32. Horst in cross section.

FIG. 8-30. Strike-slip fault. Oblique air photo of San Andreas fault, California. Distant block has moved from left to right. Fault marked by dots. (U.S. Geological Survey.)

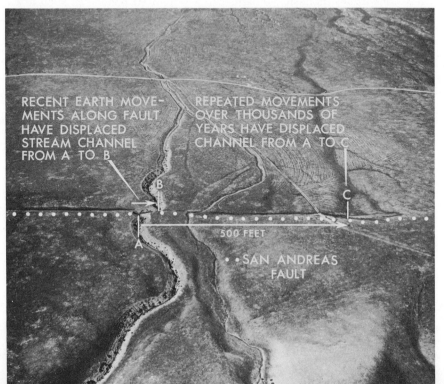

RECENT EARTH MOVE-
MENTS ALONG FAULT
HAVE DISPLACED
STREAM CHANNEL
FROM A TO B

REPEATED MOVEMENTS
OVER THOUSANDS OF
YEARS HAVE DISPLACED
CHANNEL FROM A TO C

B

C

500 FEET

A

•SAN ANDREAS
FAULT

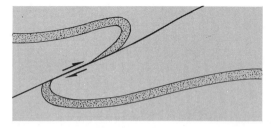

FIG. 8-33. Reverse fault and overturned fold in cross section. In a nearby section where motion is less, a fault may not exist.

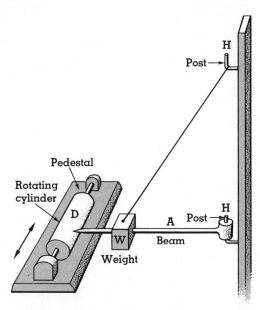

FIG. 8-34. Essential parts of a seismograph.

In any case, although they are visible at the surface, they also have a subsurface existence. Consequently an anticline is not the same thing as a hill or mountain, nor is a syncline the same as a valley. The distinction is one between the forms at depth of the geologic structure and the superficial forms of the land surface.

EARTHQUAKES AND THE EARTH'S INTERIOR

The earthquake is one of the most impressive—and frightening—of geological phenomena. The degree of devastation it brings, the large areas affected, the rapidity of the process, and the lack of warning all impress us with the power of forces we do not really understand. The side effects set off by earthquakes add to their impact—rock slides, landslides, damming of rivers often followed by disastrous floods when the dam is breached, tsunamis (so-called tidal waves), changes in the level of the land, and fire and disease that follow the initial disturbance.

Motion is the outstanding element of the earthquake—motion that is visible, in the strongest quakes, as waves on the surface of the land; that may be felt, as vibrations of the earth; and that has been detected many other times, in recent years, by the sensitive instruments that are now available. The device used to record and measure earthquake movement is the *seismograph*. Its basic elements are a weight suspended in such a way that inertia holds it still while the surroundings are in motion, and a means of recording the surrounding motion so that the direction, frequency, and amplitude of the waves can be determined. The essential parts of a seismograph are shown in Fig. 8–34. The post and pedestal are firmly seated in bedrock. The weight (W) on the arm (A) remains motionless and a penpoint on the end of the arm traces a line on the revolving drum (D) when the earthquake causes vibration of the ground parallel to the axis of the drum. Because the motion of the weight is restricted by the two hinges (H), this seismograph shows only motion in the direction of the double-headed arrow. The addition of two more seismographs and the placement of the three with axes at right angles to each other would make possible more complete data. The oscillations of the drum produced by the earthquake are recorded by deflections of the trace on the drum, and a strip of paper covering the surface of the motor-driven drum receives the trace. The record thus made is called a *seismogram*.

KINDS OF EARTHQUAKE WAVES

Our studies of the motion of earthquakes tell us that there are three kinds of waves that are produced by the disturbance and transmit its energy. *Long waves* travel close to the surface. They give us information about near-surface rocks we cannot observe. The other two types of waves pass through the depths of the earth. It is from them that we get our information about the earth's interior. *Primary waves* are *compressional* waves; they travel in the same manner as sound. A long coiled spring extended on a flat surface provides an excellent demonstration of their behavior, as shown in Fig. 8–35. In (a) the spring is relaxed, and in (b) it has been hit by a hammer, moving the end to the right, and compressing the coils in the vicinity of C. In (c) the hammer has been removed, the end has returned to the original position, the compressed section (C) has traveled to the right, and a stretched section (S) has formed following the compressed section. In (d) the compressed and stretched parts of the spring are seen to have moved farther to the right. They will continue to the right with diminishing strength until the end of the spring is reached or until they have lost energy and are so weak as to be undetectable. Primary waves can go from one side of the earth, through the center, to the opposite side. They can travel through solids, liquids, and gases. Their velocity is about 3.8 miles per second in granite near the earth's surface and increases with depth.

To illustrate the behavior of *secondary waves* we return to our coiled spring. If the end of the spring is grasped and moved rapidly sideways, the whipping motion will put the spring into a series of waves that will move along its length. These are *shear* waves; they differ from compressional waves in their velocity, which is only about three-fifths as great, and in being able to travel through solid materials only. Because primary and secondary waves move with different velocities, it is possible by record-

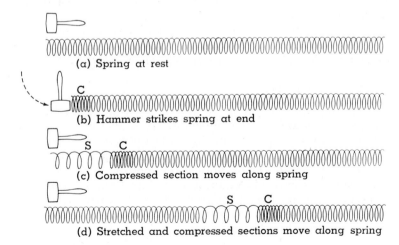

(a) Spring at rest

(b) Hammer strikes spring at end

(c) Compressed section moves along spring

(d) Stretched and compressed sections move along spring

FIG. 8-35. A compressional wave in a coil spring resting on a smooth surface.

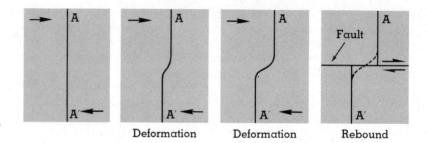

FIG. 8-36. Maps showing progressive deformation and rebound.

Deformation   Deformation   Rebound

ing the difference in their arrival times at the station to calculate the distance to the source. If the distance is determined at three different stations, the source may be located at the point of intersection of three spheres constructed with the stations as centers and the distances as radii.

TSUNAMIS

The tsunami is a sea wave produced by a nearshore or submarine earthquake, or by a volcanic eruption. (The name is from the Japanese: *tsu*, meaning port, and *nami*, wave.) The probable immediate cause of most tsunamis is some vertical motion of the sea floor.

Tsunamis usually include a moderate but abnormal rise or recession, followed by three to five major oscillations. The height may be as much as 40 feet near the source, and velocities may reach several hundred miles per hour. Even after traveling thousands of miles a tsunami may be several feet high. During the Alaska earthquake of 1964 the northern part of the Gulf of Alaska was raised (a maximum of 3.5 meters) and then drained away within 30 minutes, making a broad wave crest that spread out into the Pacific Ocean. It was 11 centimeters high at Wake Island, 5 meters at California, and 1.3 meters at Antarctica. The effects of tsunamis can be catastrophic in loss of life and destruction of property.

EARTHQUAKE ORIGIN

Many earthquakes appear to be associated with the faulting of great rock masses. It is known that immense forces may be involved in such deformations. In the diagrams of Fig. 8–36 the forces are represented by arrows and the line A-A' shows the amount and place of deformation. The rock is folded more and more, until its strength is exceeded, it ruptures, and the material almost instantaneously springs back to its original form. The vibrations accompanying the rebound are the earthquake. In the San Francisco earthquake of 1906, offset of this kind to a maximum of 21 feet occurred along a strike-slip fault.

During the 1960s many earthquake shocks originated in the vicinity of Denver, Colorado, an area previously essentially free of such disturbances. Their sudden appearance was baffling until it was discovered that their frequency and intensity could be correlated with the injection of large amounts of contaminated water into a well that reached rocks several thousand feet below the surface. The water may have lubricated faults and thus caused rock displacement, or the added pressure may have initiated motion. The discovery suggests a possible method of earthquake control: Perhaps the injection of water into bedrock in earthquake-prone areas might provide a gradual release for forces in the crust and mantle, setting off a series of small and harmless movements which would prevent the build-up of large forces of potentially greater destructiveness.

We have described two instances in which it was possible to identify the immediate cause of a seismic disturbance. But in many cases there is no discernible event that can be held responsible. The fact is that we really do not have a theory to adequately account in a general way for the change of forces and the great amounts of energy involved in earthquake upheavals. Such a theory will probably have to await the development of much fuller knowledge of the earth's interior.

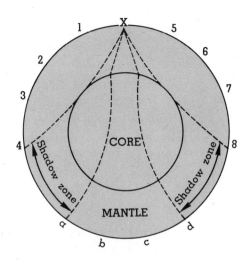

FIG. 8-37. Diagrammatic cross section of zoned earth, showing paths of refracted earthquake waves.

INTERPRETATION OF EARTHQUAKE WAVES

The source of earthquake waves—the site beneath the surface where the disturbance originates—is known as the *focus*; and the point on the earth's surface directly above, as the *epicenter*. With several seismographic stations recording the arrival times of primary and secondary waves, we can locate the epicenter of an earthquake at the intersection of circles drawn on a globe around points representing the locations of the recording stations. The depth of the focus can then be calculated. Most earthquakes have a focus in the range of 5 to 50 kilometers in depth, but some have been recorded at depths up to 720 kilometers.

The model of the earth's interior discussed in Chapter 7 was developed by studying the reception of primary and secondary waves, as shown in Fig. 8–37. From an epicenter at point X, both kinds of waves are received at stations represented by numbers. No waves are received in the ringlike zone labeled the shadow zone. Only primary waves of appreciably reduced average velocities are received in the zone a-b-c-d. An analysis of the types, distribution, and velocities of the waves, combined with the knowledge that shear waves cannot go through liquids, has established with reasonable certainty that the earth has a liquid core of radius of approximately 2200 miles and an overlying mantle about 1800 miles thick that has the properties of a solid. The fact that a central zone within the core transmits waves of higher velocities than the rest of the core suggests that the inner core is probably solid.

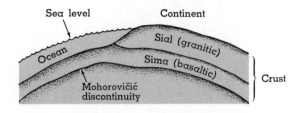

FIG. 8-38. Diagrammatic cross section of the earth's crust at the continent–ocean-basin boundary.

The properties of the materials in the crust are known from our studies of long waves, which do not penetrate more than a few miles below the earth's surface. We reach our conclusions by combining data on the velocities and paths of the long waves with our knowledge of the relation of wave velocities to rock temperatures and pressures derived from laboratory and field studies. Thus we conclude, as discussed in Chapter 7, that there exist, just beneath the surface, layers of unconsolidated sediment and sedimentary rock; next, a granitic layer under the continents 6 to 9 miles thick generally and thicker under the mountain ranges; and, finally, a basaltic layer under both continents and oceans with its base at depths ranging from 8 to 40 miles (Fig. 8–38). The granitic and basaltic layers are known as the *sial* and the *sima*, respectively. These are coined words representing the predominant elements in each: si- for silicon, -al for aluminum, and -ma for magnesium.

We are able to locate the depths at which there are changes in the composition of the earth's interior by noting where there are the sharpest changes in the velocities of the earthquake waves. The most distinct such change occurs at the depth at which we place the base of the crust. This boundary between the crust and the mantle is called the *Mohorovičić discontinuity*—more familiarly, just the *Moho*—after the Yugoslav scientist who discovered it in 1909 (Fig. 8–38). It would be of the greatest scientific interest to obtain samples of the rock above and below this boundary line, and it would be technically feasible to do so by means of oil-well drilling equipment. Such a venture was undertaken in the United States as a joint effort of scientific, commercial, and government groups, and was known as Project Mohole. Unfortunately, the project was abandoned in 1966, after the preliminary steps had been taken, because of failure to obtain additional financial support from the government. Subsequently, the Soviet Union announced a similar plan. It remains to be seen whether their project will fare any better than ours in the competition for funds with military and space programs.

THE EARTH'S INTERIOR

Materials. We are not able, as yet, to obtain samples of the materials that lie deep within the earth. But we do have indirect evidence to help us reach conclusions about their composition. It is believed that meteorites (meteors that reach the earth's surface) are composed of the same materials as the interior of the earth. They range in composition from a mixture of nickel and iron to rock-forming silicate minerals. By analyzing the components, their relative quantities, and the specific gravities, and combining these data with the information derived from earthquake waves, we are able to construct a zoned earth model (Fig. 8–39). Our model shows a planet of specific gravity increasing with depth; a crust of granitic

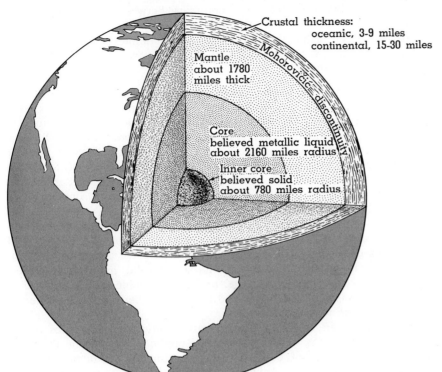

Crustal thickness:
oceanic, 3-9 miles
continental, 15-30 miles

Mohorovicic discontinuity

Mantle
about 1780
miles thick

Core
believed metallic liquid
about 2160 miles radius

Inner core
believed solid
about 780 miles radius

FIG. 8-39. Seismologist's conception of the interior structure of the earth; crustal thickness exaggerated. (After U.S. Coast and Geodetic Survey.)

and basaltic rocks, a mantle of more dense simatic rocks, and a core rich in nickel and iron; with intermediate layers of metallic sulfides or high-density silicates.

Pressure and its effects. As the depths increase from the surface of the earth to the core, the layer of overlying rock becomes thicker and the load exerted greater, so pressures within the earth are far higher than at the surface. It is estimated that the pressure in pounds per square inch approximately equals the depth in feet. The rock pressure at a depth of 1 mile would then be about 2½ tons per square inch; and you can imagine what extremely high pressures would exist in the deep interior. Geologists believe that these pressures may have the effect of altering crystal structure to such an extent that substances may have properties entirely unlike their chemical counterparts at the earth's surface. It is postulated that unit cells of a mineral can withstand increasing pressure without deformation up to a threshold, beyond which they assume a different, more compact arrangement. Thus the Moho, the boundary between the crust and the mantle, may be due to the changing pressure effects on the elastic properties of rock rather than to a change in chemical composition. We have a basis, then, for proposing that the Moho may be a physical, not compositional, discontinuity.

Temperature and its effects: the geothermal gradient. The soil and near-surface rock are heated and cooled by radiation and conduction that depend on surface conditions. But below a certain depth temperatures in the earth do not change either seasonally or in shorter cycles. Observations made below this zone of fluctuation, in mines and oil wells, indicate that temperature increases with depth about 1° F. per 60 to 100 feet. This rate of change is called the *geothermal gradient*. In volcanic areas it is greater. The greatest depth for which we have temperature data is about 24,000 feet, and this is insignificant when compared with the earth's radius of 21 million feet. So while it is reasonable to assume that temperatures continue to rise toward the core, the temperature at the center of the earth cannot be estimated with any accuracy. We are sure, however, that both the elasticity of matter and its state (solid, liquid, or gaseous) are affected by temperature change, and it is possible that the Moho or the boundary between the mantle and the core, or both, may be related to temperature belts within the earth.

The combined effects of high temperatures and high pressures in the earth's interior are as yet impossible to predict. We cannot duplicate extremely high values of both at the same time in the laboratory, and physicists have not yet been able to work out the results mathematically. In any case we do not know what temperatures actually exist at great depths. If a temperature gradient of 1° F. per 100 feet of depth is extrapolated, temperatures above the melting points of most minerals are exceeded at depths much less than that of the base of the mantle, and yet the mantle clearly has some properties of a solid. It is believed that the great pressures at these depths prevent the melting of rock and thus maintain the solid state except locally, where a higher than average content of radioactive minerals, or perhaps some other factor, may melt the rock and cause the formation of magma. Perhaps you will feel dissatisfied with the speculative nature of these conclusions, but for the time being this is as much as we can say with assurance.

THE CRUST AND MOUNTAIN MAKING

We spoke earlier of the distinction between the geologic structures that exist beneath and sometimes partially exposed at the surface and the purely surface forms. The most outstanding and perhaps most interesting of the latter are the earth's mountains, which fall into two groups. Think for a moment of the shape of the surface as the product of some unseen sculptor of vast powers. A sculptor may create his forms either by the addition of material to make a pleasing shape or by carving away surrounding material to leave the shape he envisions. Similarly, we have two kinds of mountains:

1. *Volcanic* mountains are *aggradational*. They are built up from lava, in the form of shield volcanoes, cinder cones, and composite cones.
2. *Erosional* mountains are *degradational*. They are high-standing remnants left over after surrounding materials have been removed by the action of wind, water, or ice.

We have already discussed the first group, so it is the second to which we now turn our attention. Erosional mountains are divided, again, into three types:

a. *Block faults*, formed on great blocks, some downdropped and some uplifted, and carved into mountains by erosion.
b. *Domal* mountains, formed by erosion of an area uplifted in a single, simple, large fold.
c. *Folded and faulted* mountains, so complex in form and history that we shall have to consider them in detail. These are the great mountain chains still in existence and the remnants of equally great chains that were formed many millions of years ago—a group that is worthy of special study because it is intimately related to the composition, structure, and history of the earth.

ROOTS OF MOUNTAINS AND ISOSTASY

We referred earlier to the fact that the base of the earth's crust under the continents is believed to be at depths that range from 25 to 40 miles, and that the granitic layer of the crust under continental areas is thicker beneath the mountain ranges than elsewhere. We shall now consider the uneven thickness of the crust and its relation to mountain building.

Evidence for the variable thickness of the granitic layer rests on calculations and measurements of the force of gravitational attraction at various locations on the earth's surface. Physicists tell us that all material bodies attract each other with a force that is directly proportional to their masses and inversely proportional to the square of the distance between them. (The mass of a body is an expression of the amount of matter it contains, and on earth is indicated by its weight; but our weight changes if we leave the earth's gravitational field, whereas our mass remains the same. The statement of inverse proportionality indicates that if the distance between two bodies is doubled, for example, the force of attraction between them will be divided by 2^2, or 4.) On the basis of the law of gravity we know that if the earth were a perfect sphere with its mass evenly distributed throughout, a pendulum suspended above its surface would hang

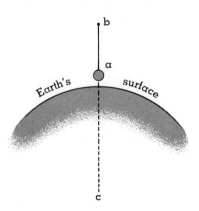

FIG. 8-40. Pendulum on a spherical homogeneous earth.

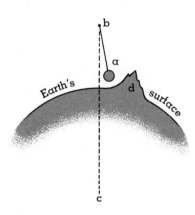

FIG. 8-41. Pendulum near a mountain range.

FIG. 8-42. Diagrammatic representations of mountain ranges, (a) as upward extensions of the crust, and (b) as blocks of low specific gravity resting on a crust of higher specific gravity.

with the center of its weight on a line with the center of the earth, as shown in Fig. 8–40. The same condition would obtain if the earth were made up of perfectly concentric layers of different composition (i.e., different specific gravity), so long as each layer was of uniform specific gravity throughout. We know, however, that our earth is not a perfect sphere, nor are all its layers perfectly concentric, for it has elevations of matter—that is, its mountains—extending above its generally almost spherical surface. Again, the law of gravity makes it possible for us to calculate on the basis of the composition, size, and shape of a given mountain range the horizontal attraction due to its mass, assuming it to be a body of material resting on the surface. We know, then, what deviation from the vertical to expect for a pendulum suspended in the vicinity of any mountain range for which we have the required information (Fig. 8–41). When such calculations were made and the related observations performed for several mountain ranges, the actual deviations were found to be less than the calculated deviations. How was this to be explained?

It was apparent that we could consider mountains neither as upward extensions of the crust, as shown in Fig. 8–42(a), nor as blocks of material of low specific gravity resting on a crust of higher specific gravity, as illustrated in Fig. 8–42(b); for neither of these models would yield the observed results. Thus geologists came to the conclusion that mountain ranges must be viewed as blocks of low specific gravity "floating" in a substratum of higher specific gravity (Fig. 8–43). The extent to which the blocks extend above and below the surface is related to their specific gravity as compared with that of the crust, and represents a state of equilibrium called *isostasy*. We know that different mountain ranges may be composed of rock types of different specific gravity, so our theory must account for mountains of similar rock but different height and mountains of the same height but different specific gravity. We have two different models to illustrate how isostasy is achieved in these circumstances. Two

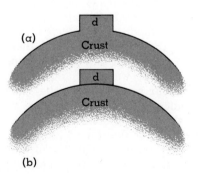

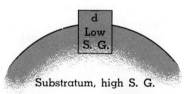

FIG. 8-43. Model of mountains as blocks of low specific gravity "floating" in crust of higher specific gravity (greatly exaggerated vertically).

British scientists, J. H. Pratt and G. B. Airy, were responsible for proposing these interpretations, and the name of each has been given to the theory he originated.

In each model the earth is represented as a liquid in which solid blocks —the mountains—are floating. The upper surfaces of the blocks represent the topographic highs of mountainous regions. In the *Pratt* theory mountains of different composition are represented as blocks of different specific gravity, with heights that differ in inverse proportion to their densities. The bases of all the blocks are at the same level, and the blocks of lowest specific gravity would be of greatest height, the blocks of highest specific gravity of least height (Fig. 8–44). This theory does not account for mountain ranges that are of different height but composed of rock of the same specific gravity.

In the *Airy* theory mountains composed of the same material are represented as blocks of the same specific gravity but different heights and with their bases extending to different levels in the denser substratum (Fig. 8–45). The bases of the highest blocks extend to the greatest depths. Bases which extend below the average level for continental areas and thus dip into the basaltic layer of the crust are referred to as the *roots* of mountains. Many mountain ranges that have been investigated, including the Himalayas, Rocky Mountains, and Sierra Nevada, have been found to have roots. These mountains are considered to be in isostatic equilibrium. The few known ranges that do not have roots are not in equilibrium and must be supported as uplifts in some manner other than by flotation in the crust and upper mantle.

Because the Airy theory has been confirmed by considerable evidence and is generally accepted, we think of the earth's crust as portrayed diagrammatically in Fig. 8–46. Note the thickening of the sial, or granitic layer, below the mountain range, and the corresponding dip in the base of the crust at the same vertical location.

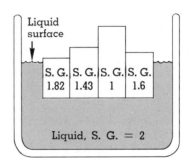

FIG. 8-44. Model illustrating isostasy according to Pratt. Blocks of different specific gravities floating in a liquid.

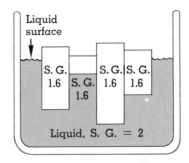

FIG. 8-45. Model illustrating isostasy according to Airy. Blocks of uniform specific gravity floating in a liquid.

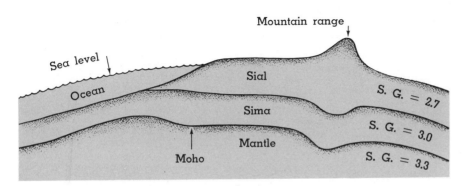

FIG. 8-46. Cross section of the crust showing differences in ocean basins, continents, and mountain ranges. Not to scale.

The great mountain chains of the world are comprised of folded and faulted mountains. Such ranges are characterized by three outstanding features: (1) deep roots; (2) a sedimentary rock layer much thicker than average; and (3) a complicated geologic history generally including events in a well-defined sequence, with isostatic adjustment and equilibrium as important features of their development. The history of such a mountain chain, much simplified and generalized, follows; but bear in mind that not all mountain ranges conform exactly to this sequence, and variations are numerous.

1. The process begins with the gradual formation of a very large elongate trough in the crust of the earth, perhaps as much as several hundred miles wide and a few thousand miles long. This depression is called a *geosyncline*. We do not know with any assurance what forces within the earth may be responsible for such deformations.

2. Sediment derived from the adjacent higher areas is transported to the geosyncline and deposited in it, usually in marine waters, to form sedimentary rock. Depression of the geosyncline is slow, and sediment is deposited at a comparable rate, so that the water is generally comparatively shallow (seldom more than a few hundred feet deep). The rocks are fossiliferous marine sediments, similar to sediments now being deposited in shallow water adjacent to continents. Basins of greater depth may exist locally, and there deep-water sediments are deposited. A more rapid rate of subsidence of the geosyncline is likely to result in deposition of a greater thickness of sediment, assuming the supply is adequate. Some extrusives and pyroclastics may also be deposited. This phase of the process may continue for millions of years, broken only intermittently and briefly by uplifts of short duration. Many thousands of feet of sediment may thus be deposited.

3. Subsidence eventually stops and a different set of forces now affects the area of the geosyncline: horizontal compression dominates, throw-

FIG. 8-47. Idealized cross section showing folds and reverse faults caused by forces directed to right.

ing the layered sedimentary rocks and the underlying materials, whether igneous, metamorphic, or sedimentary, into folds and reverse faults of great intensity. If the compressional forces at work are from one direction only, the folds may be overturned, and the amount of displacement on thrust faults will be greater in the direction of the forces (Fig. 8–47).

4. Intrusion by granitic batholiths, or granitization deep within the sedimentary strata, are now likely to occur, following which the compressional forces are relaxed.

5. At this point tension parallel to the earth's surface becomes dominant. Rocks of the crust are broken by normal faults; block faulting with step faults, horsts, and grabens is conspicuous; and extrusion of basaltic lavas may be widespread.

6. The geosyncline at this stage is a great downwarp in the crust, filled with sediments of comparatively low specific gravity. As long as compressional or other forces continue to act, isostatic equilibrium cannot be reached. But when no other forces are operating, the principle of isostasy governs and the geosynclinal body of material rises because of its low specific gravity. The folded and faulted sediments are uplifted isostatically, perhaps more rapidly than erosion can cut them down, and a belt of mountains will thus exist along the axis once occupied by the original depression. In the area of maximum uplift, granite batholiths or granitized rocks may be exposed. Because of their greater resistance to erosion, they will be found in the high mountains along the center of the range, flanked by foothills of sedimentary or less metamorphosed rock.

The Appalachian Mountains and the Alps are examples of mountain chains of the folded and faulted type. The Appalachians are much the older and are thus especially interesting, for the northeastern end has gone through the cycle we have just described three times. The Alps, with a much shorter history, have not been so reduced by erosion.

MOUNTAINS AND THE OCEAN DEEPS

Although a sequence of events culminating in isostatic equilibrium is the pattern for the history of most of the great mountain ranges of the world as far as we now know, we do have evidence of a different type of deformation in some areas, notably the islands of the East Indies and certain island chains of the West and Southwest Pacific Ocean. These island chains exist in great arcs, with the deeps of the Pacific Ocean on the convex side of the arcs. It has been found that along long, narrow belts over these deeps the force of the earth's gravity is much less than calculations would lead us to expect. A Dutch geophysicist, F. A. Vening Meinesz,

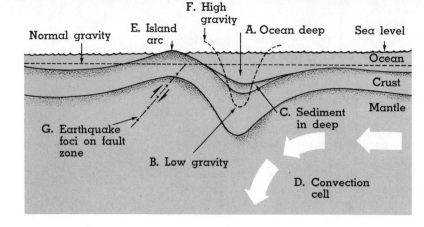

FIG. 8-48. Diagrammatic cross section of tectogene. Strength of gravity indicated by dashed line. Large arrows show convection currents in the mantle.

in the earlier part of this century proposed an explanation for this phenomenon that has been widely accepted. The fundamental elements are shown schematically in Fig. 8–48 and may be best understood by reference to the diagram:

A. The ocean deep, or *tectogene*, in which the crust bends downward, dipping into the mantle, and is also of increased thickness.
B. The low-gravity belt that follows the course of the deep.
C. The crust and sediment in the deep may be intensely folded and faulted, and perhaps metamorphosed where depressed the farthest.
D. A great convection cell in the mantle, which is thought to be the force responsible for the tectogene; the crust is deformed by the downdrag of the convection current where its motion is toward the center of the earth.*
E. The island arc, which is on an anticlinal fold of compressional origin; if only one convection cell is active, the curve of the arc indicates that the cell is beneath and on the concave side.
F. High-gravity belt over the island arc.
G. Rotation of the convection cell may cause reverse faulting, with fault surfaces on the concave side of the arc and dipping away from it; earthquake foci are abundant along such planes, with shallow foci nearest the island arc and foci of greater depth farther away (the deepest known foci—720 kilometers—have been found to be associated with such planes).

* You will recall from our discussion of atmospheric phenomena (Chapter 2) that air which is heated expands and rises over the cooler air of greater density that surrounds it, and thus forms a convection current. Other kinds of matter may also, when heated, behave in this manner. A convection current in the mantle becomes circular, in cross section, forming a *convection cell*, as the flowing rock material gradually cools on nearing the crust, levels off in its path, and then descends.

EARTH SCIENCE

If at some time the convection cell should become inactive, isostatic events would then come into play, and the tectogene would rise to form a mountain chain composed of crustal rocks and deep-water sediments metamorphosed by the high pressures and high temperatures to which they were exposed when at great depths. The existence of many mountain chains of Southeast Asia can be related to island arcs, and can be utilized to trace the positions of a succession of convection cells, island arcs, and tectogenes of the geologic past.

CAUSES OF MOUNTAIN BUILDING

In discussions of mountain building, geologists are likely to be more certain of the course of events that produced the final structures than they are in identifying the forces responsible for the original deformation. This situation reflects the sparsity of our knowledge about the interior of the earth and the forces at work there. Nevertheless, we do have some ideas on this subject, not by any means mutually exclusive. The likelihood is that if we are ever to explain adequately all the mountain ranges of the world it will only be by a combination of theories that will take into consideration the vast complexity of the earth's systems and their interaction.

Convection cell. The hypothesis of the convection cell appears to have considerable merit in explaining geological phenomena in some parts of the earth. Although we do not ordinarily think of the mantle as a fluid in which currents may flow, it is quite possible that such a condition may be brought about in given areas as a result of confining pressure exerted over long periods of time and producing plastic deformation. Where, then, rock material of the mantle exists as a liquid of extremely high viscosity, convection cells may arise consisting of a volume of rock of lower-than-average specific gravity ascending through surrounding denser material and forming a circular current. Rock of reduced specific gravity may occur through heating by locally high concentrations of radioactive elements, or by high concentrations of low-density materials such as water or other volatiles.

Continental drift. Fairly substantial evidence exists in favor of the hypothesis of an early, large, single continent, which has broken into fragments that have drifted apart. Folded and faulted mountain ranges are thought to have formed as wrinkles caused by friction at the leading edges of the drifting continents. This explanation cannot, of course, account for mountain chains in other continental locations, nor for the anomalous positions of earthquake foci, especially those associated with island arcs. The energy source for continental drift is postulated to be the convection cell.

Contracting earth. As mentioned earlier (Chapter 7), one theory of the earth's origin holds that the planet formed as a body of incandescent gas which though now largely solidified is still cooling and contracting. Such shrinkage would mean that the crust and the mantle are under compression. This could account for horizontal compressive forces as the source of crustal deformation. Aside from the fact that this theory is still very much challenged by those who believe the earth originated by slow accumulation of solid fragments, other objections may be raised to it specifically in relation to mountain building. First, mountain ranges are not as numerous over the earth as they would be if caused by the earth's contraction. Second, presently existing mountain ranges follow two major directional trends—eastward along the Alps-Caucasus-Himalaya trend and northward elsewhere in the world—whereas a more random orientation pattern would be predicted on the basis of this theory. And, finally, there is more total compression in orogenic belts than could result from thermal contraction alone.

SUGGESTED REFERENCES

Bascom, Willard: *A Hole in the Bottom of the Sea: The Story of the Mohole Project*, Doubleday & Company, Inc., Garden City, N.Y., 1961.

Eardley, A. J.: *Structural Geology of North America*, 2nd ed., Harper & Row, Publishers, Inc., New York, 1962.

Garland, G. D.: *Earth's Shape and Gravity*, Pergamon Press, Inc., New York, 1965. (Paperback.)

Hodgson, John H.: *Earthquakes and Earth Structure*, Prentice-Hall, Inc., Englewood Cliffs, N.J., 1964. (Paperback.)

Jacobs, J. A.: *Earth's Core and Geomagnetism*, Pergamon Press, Inc., New York, 1963. (Paperback.)

Lahee, F. H.: *Field Geology*, 6th ed., McGraw-Hill Book Company, Inc., New York, 1961.

Leet, D. J., and Sheldon Judson: *Physical Geology*, 3rd ed., Prentice-Hall, Inc., Englewood Cliffs, N.J., 1965.

Rittman, A.: *Volcanoes and Their Activity*, John Wiley & Sons, Inc., New York, 1962.

Rogers, John J. W., and John A. S. Adams: *Fundamentals of Geology*, Harper & Row, Publishers, Inc., New York, 1966.

ROCK WEATHERING,
SOIL FORMATION,
AND GEOLOGIC TIME

CHAPTER 9

9

Weathering is the process by which hard, strong rock is changed in the surface or near-surface environment by physical or chemical means to a granular form, or to rock substantially less resistant to the action of the elements. Just as a sculptor can more rapidly carve a soft rock than a hard, dense one, so can landform sculpture occur more rapidly and with less expenditure of energy if soil or rock fragments are involved rather than large rock bodies. Weathering may thus be considered the precursor of the landforming processes.

Some of the results of the weathering process are familiar—the green patina that forms on copper exposed to the atmosphere, the rusting of unprotected iron and steel in a humid climate, and the loss of clear-cut detail on gravestones. Weathering phenomena fall into two groups: those consisting of mechanical action and those that involve chemical change. This division, like many others we have studied, is not an absolute one, and the categories sometimes overlap. Either type of weathering may occur anywhere from the surface to a depth of 400 feet, depending on the climate and the properties of the material being weathered.

The geologist is interested in weathering because it is an aspect of the history of the earth and the development of landforms, but the process has practical importance as well. The agronomist is concerned because weathering affects soil, the civil engineer because it determines the properties of many foundation and construction materials. The products of weathering enter our everyday lives to a startling extent: a well-known geologist in the United States recently pointed out that an average serving of beefsteak contains elements derived from the weathering of two ounces of rock; the paper of this book is coated with clay formed in the weathering process; and some uranium minerals used in nuclear plants for the generation of electric power are weathering products.

MECHANICAL WEATHERING

Mechanical weathering (also called *disintegration*) produces a change from a few large rock bodies with a small total surface area to many small fragments with a large total surface area (Fig. 9–1). The small fragments may then be more easily transported by water, wind, or ice, and the increase in surface area facilitates further weathering by chemical action.

Frost wedging and *plant wedging* are two of the most important methods by which rocks may be fragmented. Wedging by frost depends on the fact that water expands in volume by about one-ninth when it changes from liquid to solid state. Where rock joints, spaces between cemented clastic grains, or pore spaces in scoria or pumice are more than 91 per

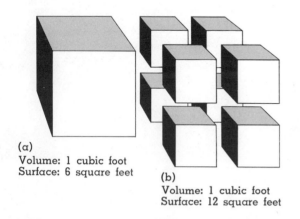

(a)
Volume: 1 cubic foot
Surface: 6 square feet

(b)
Volume: 1 cubic foot
Surface: 12 square feet

FIG. 9-1. Comparison of surface areas for one large cube and a cube of the same size cut into eight small cubes.

cent full of water, the expansion of freezing will produce enough internal pressure to cause rupture. Laboratory tests have confirmed that pressures can thus be created that are much greater than the strength of any rock to resist. Fragmentation by frost can, of course, take place only in those parts of the world where water is present and is frozen part of the time. In tropical regions, where water never freezes, and in polar regions, where it never thaws, frost wedging is inoperative. At intermediate latitudes, and high altitudes at lower latitudes, much frost wedging occurs. On steep, bare mountain slopes where there are daily freeze-thaw cycles, the first warming by the sun in the morning brings a constant clatter of particles separated by nighttime freezing, released by the thaw, and rolling down the slope.

Plant wedging operates to enlarge rock spaces through the pressure of plant roots growing in them. Eventually, complete rupture occurs along partly opened fractures.

CHEMICAL WEATHERING

Chemical weathering (also called *decomposition*) generally, but with one important exception, involves the formation of new minerals through the chemical combination of preexisting substances. Like other chemical processes, it is most rapid at high temperatures and where there is the greatest degree of contact between the reacting materials. The possibility of substantial contact is often created through an increase in surface area brought about by mechanical weathering; thus the two kinds of weathering frequently occur as a sequence in time.

Let us first take note of the important exception mentioned above before we go on to more typical chemical weathering phenomena. *Solution* depends on the fact that most minerals are, to a greater or lesser extent, soluble in water. You can imagine, then, that it is a widespread occurrence. Rainwater falling on bare granite and flowing only a few feet has been found to have dissolved material in measurable amounts. Note however, that solution does not involve a change of chemical composition. Thus calcite, for example, does not become a different mineral substance when transported in solution. The process of solution plays an important role in enlarging preexisting openings in rock bodies and in transporting materials to areas where they have not previously existed.

Hydration is the process by which water (H_2O) combines chemically with a mineral to form a new mineral. Hydrated minerals are of lower specific gravity and softer than the original mineral, and are often different in other respects. Anhydrite ($CaSO_4$), for example, hydrates to form gypsum ($CaSO_4 \cdot 2H_2O$) and expands in the process. In unlined tunnels carrying irrigation water, anhydrite has been known to expand

FIG. 9-2. Early stage in plant succession and weathering process. Lichens, mosses, and small ferns growing on a gabbro boulder. (U.S. Department of Agriculture.)

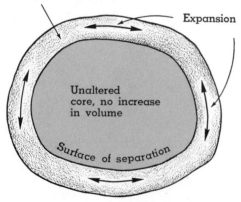

Weathering and expanding layer near surface

Expansion

Unaltered core, no increase in volume

Surface of separation

FIG. 9-3. Exfoliation of a boulder.

so rapidly that within a few years excavation has been required in order to restore the tunnel to its original diameter.

Reactions involving water, carbon dioxide or oxygen, and a mineral are perhaps the most important of the chemical weathering processes. The feldspars, common igneous rock minerals, are often involved in such reactions. One of the plagioclase feldspars, for example, reacts with water and carbon dioxide as indicated by the following chemical equation:

$$2NaAlSi_3O_8 + 2H_2O + CO_2 \longrightarrow Na_2CO_3 + 4SiO_2 + Al_2Si_2O_5(OH)_4$$

feldspar   water   carbon dioxide   sodium carbonate   quartz   clay

The process may be direct, as shown here, or it may involve a series of intermediate steps with intermediate products before the end result is reached. Other minerals of the igneous and metamorphic rocks take part in similar reactions, but with different end products. Minerals that contain a high percentage of iron release a compound of iron, such as limonite, as a product of hydration. Thus rocks bearing biotite or other iron-rich minerals should not be used for exterior trim on buildings in a humid climate because the limonite formed in weathering stains the rock an unattractive rusty color.

Soluble products of hydration are for the most part carried away, and eventually reach the oceans as a final collecting place, where most become parts of sedimentary rocks.

Plants may accelerate weathering processes by withdrawing dissolved materials and contributing organic matter, both of which tend to increase soil acidity and the rate of weathering. An early stage in plant weathering is illustrated in Fig. 9–2, which shows lichens and mosses growing on bare rock. Alfalfa grown as a hay crop derives its nutrients from the equivalent of over 2 tons of rock weathered per acre per year.

Exfoliation is a weathering process in part mechanical and in part chemical. It involves unjointed rock, ranging from boulders up to bodies several acres in area, which are altered in concentric layers parallel to the external surface, as illustrated in Fig. 9–3. The outermost layer expands as a result of formation of new minerals by hydration and consequently separates from the unaltered rock in the interior. This process is repeated many times, with deeper and deeper layers separating, until the structure of the rock resembles that of an onion. Layers may be only a few inches thick, as on small boulders, or several feet thick on such large exposures as Stone Mountain, Georgia, and Half Dome and Liberty Cap in Yosemite National Park (Fig. 9–4). The separated layers may break up, leaving curved fragments in a heap surrounding the remaining unweathered rock. It had been thought that temperature changes from day to night might cause expansion and contraction and might thus be responsible for such flaking of rock, but exhaustive studies indicate that temperature alone would not produce this effect.

FIG. 9-4. Exfoliation of a massive outcropping of igneous rock, Liberty Cap, Yosemite National Park. Note parallelism between surface and the thick separated layers. (National Park Service.)

We know that some minerals are more prone to participate in chemical reactions than others. This susceptibility to chemical change may be understood in terms very similar to those we used in discussing the vulnerability of rocks to metamorphic change (Chapter 8). Recall two concepts that are fundamental to an understanding of earth science: "environment" and "stability." Let us review them in detail now, for as you see they have application to more than one area of study.

Environment is the sum of all the conditions—physical, chemical, and biological—in which a material or an organism exists. *Stability* may be defined as the state in which a material or organism is so suited to its environment that there is no tendency for it to be altered by, nor to alter, the environment. *Instability* entails the incompatibility of a material or organism with its environment, and is thus the instigator of change. We have seen that some of the minerals of the igneous rocks crystallize out at very high temperatures and pressures and others only after temperature, and sometimes pressure, has decreased; and that this occurs in a recognized sequence (Bowen reaction series). Because earth-surface conditions entail much lower temperatures and pressures than the environment of formation of any of the igneous rock minerals, these minerals are all unsuited to surface conditions to one degree or another and are therefore all unstable in the surface environment. The first-formed igneous minerals are the *least* suited to and the *least* stable at the surface. Consequently they weather the most rapidly, and the sequence of weathering rates that has been recognized closely follows the sequence of mineral formation in igneous rocks. Muscovite and quartz crystallize late, hence they are likely to be found unaltered in rocks and clastic sediments; olivine and pyroxene crystallize early in the cooling sequence, are comparatively unstable at earth-surface conditions, and are generally found on an outcrop in substantially altered form. As a further illustration of differences in stability, igneous boulders of any rock type are more likely to show exfoliation in a humid climate than are limestone boulders, which were formed at earth-surface conditions.

SOIL FORMATION

The word soil has a number of meanings depending upon who uses it. To the construction engineer soil means the uncemented materials that can be handled with pick, shovel, and wheelbarrow, and that cover bedrock. To the farmer it is the uppermost few inches or feet of granular material that supports crops. We shall use the earth scientist's definition of soil: the near-surface fragmental material altered by the characteristic weathering process of the environment so as to form distinct layers parallel to the surface. This material is primarily a mixture of mineral grains (from clay-size to sand-size) derived from rock weathering and more-or-less finely divided organic matter. A prismatic structure perpendicular to the surface is present in many soils.

An early and widely used classification of soils was based on whether the soil was residual and formed by weathering of bedrock in place or the material from which the soil was formed was transported from elsewhere. Studies by Russian soil scientists, the first to attack the problem of soil formation scientifically and in depth, indicate that this basis of soil origin and classification is not valid.

CONTROLS OF SOIL FORMATION

Five significant controls of soil formation are now recognized. The first is *climate*; as the determinant of temperature and moisture conditions, it is decisive in controlling whether physical or chemical weathering will dominate and which particular forms of weathering will operate. Climate regulates chemical reactions so that a particular group of reacting primary minerals will combine differently in different climatic regions to produce unlike sets of secondary minerals.

The amount and frequency of precipitation is of decisive significance, since water is involved in both chemical weathering (e.g., hydration and solution) and mechanical weathering (frost wedging). Water is also responsible for most transport of solid and dissolved matter in the soil. Wet-dry cycles encourage alterations in the soil and provide removal of gases and their return, or their replacement by others. Extremes of temperatures are also important, for temperatures below freezing cause ice to form, and high temperatures and low humidity produce drying of the soil. Both temperature and moisture conditions influence the kinds of living organisms found.

A second control of soil formation is the biota, the complete animal and vegetable population that exists where soil is forming. Plants affect soil development by opening channels for passage of air and water through root penetration and subsequent decay. They add organic matter to the soil when roots die in place, or by accumulation of leaves, stems, bark, and wood above the inorganic soil. The shading they provide controls the amount of sunlight striking the soil and thus aids in determining moisture content.

When plant tissue decomposes, it forms active organic acids and carbon dioxide. As water percolates down through the soil it carries with it the finely divided organic matter and dissolved acids, carbon dioxide, and oxygen; as a result, chemical action is not restricted to the surface but goes on at some depth. Microscopic plants and animals live throughout a substantial thickness of the soil and participate in the soil-forming process by making available oxygen and carbon dioxide, sometimes in places they would otherwise not be present.

Burrowing types of animals open passageways for oxygen, carbon dioxide, and water to substantial depths. Earthworms bring to the surface appreciable amounts of inorganic and organic matter, and mix it with surface materials. The same mixing function is performed by larger animals, both because they actually carry excavated material to the surface and also because their abandoned burrows are likely to be filled with washed-in surface-derived material. There are some species of worms that take soil into their systems, extract the available nutrients, and excrete the undigested portion, thus physically and/or chemically altering the inorganic grains.

Slope is a third control of soil formation. Hilltops are likely to be well drained and are therefore conducive to a different biota than are the lower slopes and depressions, which tend to have a higher and more constant moisture content. Similarly, steep slopes are better drained than gentle ones. Soil moves down steep slopes faster, and mixing may encourage weathering but will retard the development of the *horizons*, or differentiated layers of the soil. The degree of drainage further determines the number and efficiency of the freeze-thaw and wet-dry cycles.

The angle of slope and the direction the surface faces have much to do with the microclimate. The more nearly perpendicular the slope to the average inclination of the sun's rays, the greater the amount of solar energy received. A very steep, north-facing slope in the Northern Hemisphere may never be in sunlight, whereas a more gently sloping, south-facing slope in the same area will receive a maximum of solar energy. Soil temperature, moisture, evaporation, and the concomitant physical, chemical, and biological processes and products all vary with the angle and orientation of the slope. Such differences are most conspicuous in mountainous regions, where there are extremes of slope, but they also occur elsewhere in less obvious form.

The *length of time* that weathering and the other soil-forming processes have continued is the fourth significant factor in soil formation. In an unchanging environment, the longer the processes have acted (before equilibrium is reached), the more clearly defined and the thicker the layers of the soil. The faster the processes act and the more favorable the environment for soil formation, the less the time required for establishing the layers. Thus more time would be necessary in an arctic than a tropical climate. A soil in which the layers are absent or poorly defined is called an *immature* soil, one in which the zones are all present and distinct a *mature* soil. In some areas of the world soils have been known to become layered in materials disturbed by man (such as graves or mounds) within a few centuries.

The last control, and one which, perhaps surprisingly, is strong only when all others are weak, is the *parent material*, the original material that is acted upon and eventually becomes soil. An immature soil has a substantial similarity to the parent material; but as time goes on and the primary controls act, the inorganic content inherited from the parent material is more and more modified until the end product is reached. One of the most far-reaching contributions of the Russian soil scientists is the idea that the properties of a mature soil do not depend on the parent material but on the other four controls of the soil-forming process. The more-stable parent materials will require a longer time to be reduced to mature soil but will nevertheless be altered to soil exactly like that formed from a less stable parent material in the same environment. Field observations indicate that in arid regions where relief is high and slopes comparatively steep, soils are very similar to their parent material and can be used as a guide to the bedrock materials; there the soils are immature. In warmer, more humid regions, similarities between parent material and soil are blurred.

In many parts of the world, climate and biota appear to be the dominant factors in soil formation and development, but this is not clearly so in other areas. Much work needs still to be done in working out the relative significance of the various controls.

THE SOIL—FORMING PROCESS

The essential steps in the development of a soil can be observed in an area where a parent material is newly exposed, as in a just-abandoned gravel pit. First, sparse vegetative cover consisting of a few species of hardy plants is established. As the annuals die and leaves fall, a surface layer of organic debris accumulates, then decomposes, and the decomposition products and fine fragments of organic matter are carried downward by water infiltrating the parent material. Physical and chemical weathering act on the near-surface parent material, breaking up the particles and forming new minerals that occur in extremely small units or in solution. Many of the new mineral particles, together with the finely divided organic matter, are carried downward by water and are deposited at some depth in the spaces between the grains. In an area with an excess of precipitation, most of the soluble minerals are removed. Some inorganic material, substances that are stable under these conditions, remains as a residuum in the soil near the surface. Because the physical, chemical, and biological environment at depth is different, the near-surface processes of decomposition and solution do not act there but are replaced by deposition of part of the material derived from the uppermost zone. As the processes continue, the zones become more and more distinct and the soil eventually reaches maturity.

FIG. 9-5. Temperate humid climate forest soil, showing dark organic layer (A_1) and light leached layer (A_2) above zone of accumulation (B). Scale in feet. (Soil Conservation Service.)

TABLE 9-1   ZONES OF A MATURE SOIL PROFILE IN A HUMID TEMPERATE FOREST

IDENTIFICATION	DOMINANT PROCESSES	DOMINANT CONSTITUENTS
Soil surface		
A zone	Weathering, solution, loss of dissolved and finely divided particulate material (leached zone)	Organic matter and stable residual minerals
B zone	Deposition of materials from A zone	Stable minerals derived from processes active in A zone
C zone	Mechanical weathering	Partly altered bedrock or transported materials such as sand or gravel
D zone	Too deep for activity	Unaltered bedrock

A mature soil can be divided into three zones, as shown in Table 9–1; unaltered bedrock below the C zone may be referred to as the D zone. The zones are collectively termed the *soil profile*. Soils of the same zone may differ in appearance from one area to another depending on the controlling factors. In a humid forest environment such as exists between latitudes 50° north and 60° north the organic matter is very coarse, and most is therefore in the litter on the forest floor, little penetrating deeper because of its coarseness. Precipitation is high, the mean annual temperature is comparatively low, evaporation is slow, forests are the typical vegetation rather than grass, and since the ground is shaded the litter is seldom dry. Fungi are abundant and the soil is very acid. Finely divided organic matter and some mineral grains dominate in the upper part of the A zone. *Leaching*—the removal of soluble minerals by the downward movement of water—is intense, and the residual minerals, light gray in color, are conspicuous in the A zone below the dark organic layer. Iron oxide and clay from the leached zone are transported downward and deposited in the B zone, forming a partially cemented clayey layer called a *claypan*. In some locations the B zone prevents downward movement of water, and the soil is poorly drained. The A zone may be 2 to 8 inches thick and the B zone 4 to 24 inches thick depending on the grain size of the parent material and other controls (Fig. 9–5).

In the prairie grasslands of the Corn Belt of the United States, the soil profile is substantially different. The precipitation is less, temperatures are higher, evaporation is more significant, forests do not grow, and grass is the dominant vegetation. Grass stems and leaves are fine and fragile, they decompose to finer particles than forest litter, and organic matter is carried to greater depths by downward-percolating water. Sunlight reaches the surface, the litter is drier, and it is so rapidly oxidized and decomposed that little remains. As a result there are fewer fungi, conditions are less acid, and the leaching process is not so intense. The litter is very thin; the organic layer may be 4 feet thick (although not very dark gray, since the organic matter is distributed through a great thickness); the leached zone is in many places only slightly lighter in color than unaltered parent material; and the *B* zone is the site of comparatively little deposition. *A* and *B* zones may be several feet thick. Thinner organic *A* and *B* zones of a more arid grassland soil are shown in Fig. 9–6.

In arid regions evaporation exceeds precipitation and soluble materials accumulate in the soil rather than being removed by solution. The most characteristic zone of such a soil, which is called *caliche*, is a layer cemented by soluble salts that occurs at the top of zone *C*. Caliche is formed in the following manner. Water that penetrates the soil moves downward, transporting dissolved mineral materials. After the rain stops, infiltration stops. Water in the soil may then evaporate, depositing the dissolved material as a coating on the surfaces of the soil grains and cementing them together in the *B* zone. Capillary action may lift water from greater depth, bringing with it dissolved minerals that are deposited near the surface by evaporation. A balance is established between the two processes, and caliche is formed as the result of upward motion of mineral matter in solution. In most instances the cementing substance is calcite, but any soluble salt may substitute. In the United States, caliche ranges from a thin coating on the prismatic surfaces at a depth of a few feet in central Nebraska to a layer a few inches below the surface in the arid Southwest, where it may be so well cemented that excavations can be made only by drilling and blasting. Because leaching is not far advanced, such soils can be extremely fertile if the parent material is heterogeneous—unless the soluble salt is so concentrated that it inhibits plant growth.

In a humid tropical climate the soil has still a different character. Although a great amount of organic matter may be formed per year, it decomposes rapidly and little remains in the soil. Leaching is extreme and minerals that are stable under other surface conditions, such as quartz and clays, are dissolved or decomposed and removed, leaving a predominance of oxides and hydrous oxides of iron and aluminum. If iron is present, the soil is colored red or orange-red. The end result of this process is called *laterite*. Leaching continues to great depths, in some places to

FIG. 9-6. Soil profile, chernozem, North Dakota. Note thick dark-gray *A* zone. (U.S. Department of Agriculture.)

several hundred feet. Because of the intensity of leaching, laterites are infertile. The tremendous amount of vegetation in the tropical rainforest represents nutrients that have been collected by plants over a long time and are continuously being recycled. Laterites may be so rich in iron or aluminum as to be an ore of the metals. Present economic sources of aluminum ore are laterites formed in the geologic past.

GEOLOGIC TIME

When earth scientists first attempted to connect recognized geologic agents with visible landscape features, some relationships were much better understood than others. Vulcanism was a rapid phenomenon and was clearly seen to be responsible for the growth of some mountains. Earthquakes could with certainty be associated with some uplifts and depressions of the land. But one common geologic association—a stream flowing in a valley—was for a long time not understood, posing a problem analogous to the question, Which came first, the chicken or the egg? In fact, an early idea was that a valley was made first and the stream then flowed down it. To stand on the brink of the Grand Canyon of the Colorado, to look at the tremendous chasm and then see the tiny ribbon of water (visible only intermittently along its course) that is the Colorado River, one might certainly doubt that such a puny thing as the river could be responsible for the excavation of the great volume of rock represented by the canyon. We balance the changes visible to us in our lifetime against the large size of most landscape features and find it difficult to believe that the obviously slow-acting natural processes could accomplish such great work. It is no wonder the early geologists believed in catastrophes and cataclysmic events! Even now, when we hear that erosion, so publicized by conservationists, is removing the soil of the United States at the rate of a layer 1 inch thick every 750 to 1000 years, it is difficult for us to realize how much can be accomplished by such a slow process.

When the conclusion was reached that streams excavate their valleys, the next step was to determine the length of time required. Geologists realized that the time must be very long in view of the slowness of the process. Here the scientist found himself in a dialogue with the theologian, because it was in the context of religion that time, and particularly the time since the earth, man, or deities first existed, had been considered. A great range of religious opinions has existed as to the duration of time. Certain Brahmins believed there never was a beginning of time and there never would be an end, that time is eternal. One Middle Eastern religious group held that 2 million years was the total duration of time. Zoroastrians put the total length of time at 12,000 years, and in the seventeenth century an English churchman stated that since the Creation only about 6000 years had passed.

EARTH AGE BY SEDIMENTATION

Solutions to the problem of the age of the earth were sought by geologists. It was recognized that sediment was being deposited on the bottom of the ocean. The total thickness of sedimentary rock divided by the yearly rate of accumulation should give the length of time since the deposition of sediment began. Using the best data available, an earth age of 100 million years was determined. This was so much greater than previous estimates that it was accepted for some time. But the possibility of errors eventually became apparent. The total thickness of sediment is difficult to ascertain. In some places on the continents there is no sediment, in other places upwards of 100,000 feet, and the average lies somewhere in between. Which number should be used? Other sources of error include the interruption of sedimentation, the difficulty of measuring sediment thickness on the ocean bottom, the lack of uniformity in the rate of sedimentation at all times and places, the fact that much sediment has been reworked, and the probability that there was some unknown length of time prior to the beginning of sediment deposition.

EARTH AGE BY SALT IN THE SEA

Recognizing the deficiencies in the sedimentation method, geologists looked elsewhere for a solution. In the weathering process, certain water-soluble compounds are formed and are transported to the ocean, the ultimate reservoir of these substances. One of the most soluble is halite ($NaCl$). Since the oceans are not saturated with it, presumably all the salt that has ever been formed is still dissolved there. The amount can be determined with reasonable accuracy because the general shapes and sizes of the ocean basins are known and the salt content has been determined by analysis of the water. The rate of addition of salt to the oceans can be calculated by measuring the flow rates of rivers and multiplying by the salt content of their waters. Division of the amount of salt in the sea by the annual rate of accumulation should presumably give the age of the oceans, and this figure is about 250 million years. But certain errors are inherent in the method. Large amounts of salt are known to exist as sedimentary beds, and it is possible that other deposits are as yet undiscovered. Some salt is now being reworked and this is likely to have occurred in the past. The rate of addition of salt to the sea, like the rate of sedimentation, is not constant, and some evidence indicates that both rates are higher now than their averages for the total history of the earth. In addition, the earth probably existed for an indefinite time before the process began. Note that all the sources of error would tend to make the age of the earth greater than calculated.

EARTH AGE BY COOLING RATE

The earth is known to be radiating heat. The time required to reach present conditions can be estimated by making certain assumptions and engineering calculations. The various figures obtained by this method have been substantially less than those derived by other methods, and because of a number of weaknesses in the procedure, they are not taken seriously. We do not know the historic rate and kind of heat loss, nor even whether the earth is cooling off or warming up.

EARTH AGE BY RADIOACTIVITY

The discovery of radioactivity in the nineteenth century provided the basis for a much more accurate method of determining the ages of geologic materials. Radioactive elements decay (break down) spontaneously, emitting various kinds of radiation and forming other elements that may themselves be radioactive. Ultimately, only a stable, nonradioactive element remains and the sequence ends. Investigations indicate that no external environmental conditions affect the rate of radioactive breakdown—whether temperature, pressure, combination with other elements, surface area, associated materials, or any conceivable aspect of the physical, chemical, or biological setting. The rate of decay is a

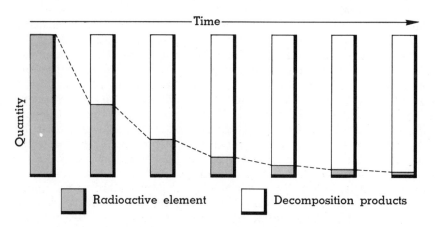

FIG. 9-7. Graphic representation of half-life of a radioactive element.

statistical phenomenon: given a quantity of a particular radioactive element, half will decay, forming the decomposition products, within a specific time called the *half-life*; half of the remaining quantity will decompose in the next such interval; half the remaining amount in the next, and so on. The process is shown graphically in Fig. 9–7.

If one knows the half-life, one need only make a chemical analysis of a mineral containing a radioactive element and perform a simple mathematical calculation to determine the time since the radioactive material was stabilized in the mineral. Caution is necessary in interpreting results, however, for some difficulties do occur. Very early in the process and very late, samples of the mineral contain very small amounts of the decay product and of the original element, respectively, and methods of chemical analysis may not be sufficiently sensitive to determine these amounts accurately. Further, if the elements have been either removed or augmented by some natural process, as may occur, the proportions in the sample will not truly represent the age of the material. Half-lives that are exceedingly long or short compared with the period of time involved cause problems in analysis. Some decay products have other, nonradioactive origins, and contamination may occur. The conditions of stabilization of radioactive minerals are also important. Most radioactive elements occur in minerals of igneous origin; age determination of a dike, for example, tells only its age and not that of the rock it intrudes. The intruded rock is older, but by how much (Fig. 9–8)? Similar problems exist with other modes of occurrence. Thus although several thousands of age determinations have been made and their accuracy fairly well established, often nearby materials can be described only as "over 3 billion years old," or "less than 750 million years old," or "between 40 million and 60 million years old."

The naturally occurring radioactive elements of greatest use, their isotope numbers, half-lives, and end products are shown in Table 9–2.

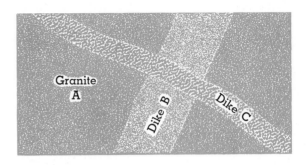

FIG. 9-8. Age relationships, in cross section. Dike B is older than dike C and younger than intruded rock (A).

TABLE 9–2 ELEMENTS OF GREATEST USE IN RADIOACTIVE DATING

RADIOACTIVE ISOTOPE	HALF-LIFE	END PRODUCT
Uranium 238	4.5 billion years	Lead 206 and helium 4
Uranium 235	710 million years	Lead 207 and helium 4
Thorium 232	15 billion years	Lead 208 and helium 4
Rubidium 87	50 billion years	Strontium 87
Potassium 40	1.3 billion years	Argon 40 and calcium 40
Carbon 14	5570 years	Nitrogen 14

From the half-lives it is apparent that only carbon 14 can be used for the youngest materials; but it cannot be used accurately for ages in excess of 30,000 years. The potassium-argon relationship is promising, since potassium is found in muscovite and some feldspars, minerals which are common in igneous rocks.

Many determinations of igneous rock age have been made. Until recently no earth materials older than 3 billion years have been found, but meteoritic substances 4.5 billion years old were known. Experience indicates that we can expect occasionally to find rocks of greater age than those previously dated, and any revisions we may make in our notions about the age of the earth will be in the direction of a greater, rather than a lesser, figure.

When we consider the great length of geologic time that has passed, we can look at the processes acting on the earth's surface and below it and at the features of the landscape and conclude that a process that appears to be slow-acting can indeed produce large changes in topography. In the next few chapters we shall look at some of these processes and the ways in which they do their work.

SUGGESTED REFERENCES

Emmons, W. H., et al.: *Geology, Principles and Processes,* McGraw-Hill Book Company, Inc., New York, 1960.

Jenny, Hans: *Factors of Soil Formation,* McGraw-Hill Book Company, Inc., New York, 1941.

Keller, W. D.: *The Principles of Chemical Weathering,* Lucas Bros., Columbia, Mo., 1959.

Leet, D. J., and Sheldon Judson: *Physical Geology,* 3rd ed., Prentice-Hall, Inc., Englewood Cliffs, N.J., 1965.

Lotkowski, Wladyslaw: *The Soil,* Educational Methods, Inc., Chicago, 1966.

Putnam, William C.: *Geology,* Oxford University Press, New York, 1964.

Reiche, Parry: *A Survey of Weathering Processes and Products,* University of New Mexico Pub. in Geology, Albuquerque, N.M., 1950.

Shelton, John S.: *Geology Illustrated,* W. H. Freeman & Co., San Francisco, 1966.

MASS WASTING, STREAMS, AND GROUND WATER

CHAPTER **10**

10

In earlier chapters we discussed the materials of the solid earth and the processes involved in their formation and accumulation; the weathering and soil-forming processes by which rock is reduced to a form that is more readily shaped by the various kinds of energy available at the earth's surface; and the aggradational landforms of volcanic origin. Most of the shapes of the landscape, however, are degradational—that is, they are remnants left at the scene after the various geologic agents have removed material by erosion. So we turn our attention now and in the next chapters to the degradational landforming processes.

Mass wasting

The spectacular landslide that occasionally halts traffic on coastal highways in California, and the "mud avalanche" that flows down a suburban Los Angeles gulch lined with homes, filling cars and houses with mud, are examples of mass movement in its most dramatic form. As a geologic agent that affects our lives, mass movement is sporadic, but its impact on landscape evolution is vastly greater, for it acts nearly everywhere all the time. The more we learn about it, the more it appears to be involved in sculpturing the land surface, to a far greater degree than we had previously believed. Mass wasting is facilitated by the deformation of rock, the weathering process, and the presence of water, acting in combination with omnipresent gravity.

DEFINITION AND SOURCE OF MOTION

Mass wasting may be defined as the downslope movement of solid and viscous materials under the influence of gravity. Water is involved in most cases, and its presence has two effects: (1) Water increases the weight of the body of material by filling spaces between grains, in pores, or in opened joints. Mass movement occurs because the downward force of gravity acting on a material exceeds its strength—i.e., its ability to resist a force—and an increase in weight may result in an imbalance that will initiate movement. (2) Many materials are less strong when wet than when dry. Dry, hard clay has substantial strength, but clay when wet and plastic will flow if a force is applied to it. In completely dry sand, slopes form at the angle of repose or smaller angles; in damp sand, the surface tension of the water film on the grain surfaces acts to hold them together; but if water is added to the point of near-saturation, the sand will flow as a thick liquid.

Materials may be near the point of movement but remain in place until some kind of triggering agent acts upon them. Once initiated, motion continues because the gravitational force exceeds the resistance. Several initiating agents have been identified. Earthquakes can upset an existing equilibrium and set off mass movement, as occurred in the Hebgen Lake, Montana, slide of 1959. Vibrations caused by heavy traffic or explosions may act in a similar manner. The thawing of frozen sediment more quickly than the water can drain out or the undercutting of a bank by stream erosion or by artificial excavation may bring about a decisive reduction in strength. Rapid wetting and saturation by heavy rain or accelerated snowmelt may increase weight and thus provide the triggering factor.

CLASSIFICATION OF MASS MOVEMENT

Classifications of mass movement have been established according to different criteria and for different purposes. We will use the criteria most commonly involved: speed of movement and properties of the material in motion.

Creep is the slowest type of mass movement; it acts on most slopes, and may be caused by a variety of agents, the most important of which are the following:

1. Freeze-thaw cycles displace soil, as shown in Fig. 10–1. Water in the soil freezes to form ice, which becomes concentrated in lenses and layers roughly parallel to the surface. As the ice mass expands, it pushes the overlying soil toward the surface in the direction perpendicular to the slope of the ground surface, as indicated by the solid arrows. When the ice melts it no longer supports the overlying soil, which is pulled vertically downward by gravity, indicated by the broken arrows. Repeated movement on this zig-zag path brings about a generally downslope motion of the soil.
2. Some clays expand when moistened and contract when drying, resulting in similar downslope movement.
3. Root growth forces soil to move laterally, and gravity aids by preferentially moving the soil particles downslope more than upslope.
4. Burrowing animals have somewhat the same effect when they enlarge their tunnels by pressing against the walls; and they contribute further to downslope movement by piling excavated material on the downslope side of the burrow entrance.
5. Larger animals displace the soil downslope wherever they step.

Most of these agents have their greatest effect near the surface; the downslope speed of soil is therefore greatest at the surface and decreases with depth. The result is not only to move soil from higher to lower elevations but also to tilt solid objects that are embedded in the soil. That is

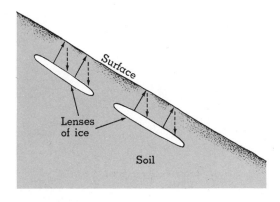

FIG. 10-1. Motion of freeze-thaw–actuated particles in creep.

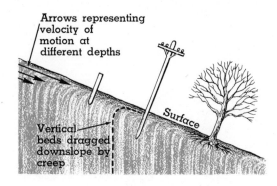

Arrows representing velocity of motion at different depths

Surface

Vertical beds dragged downslope by creep

FIG. 10-2. Cross section showing creep.

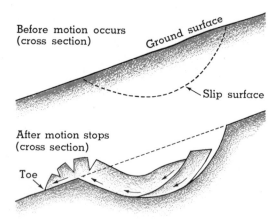

Before motion occurs (cross section)

Ground surface

Slip surface

After motion stops (cross section)

Toe

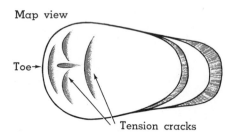

Map view

Toe

Tension cracks

FIG. 10-3. Cross sections and map view of an earthflow.

why we so often see fence posts, gravestones, and telephone and electric transmission line poles that lean downhill. Trees may also be tilted and because of their growth pattern may have curved trunks. Linear structures such as fences and roads may be offset in the downhill direction by an uneven duration or velocity of movement. In some exposed cross sections, marker beds may be seen which show the downslope movement. Figure 10–2 illustrates many of these phenomena.

Creep acts so slowly that the motion is imperceptible; but because the disturbing agents are so widespread and gravity is omnipresent, the total daily movement on all the slopes of the world must be immense.

Earthflow is a more rapid type of mass movement, in which the unit in motion is comparatively large and flows partly as a plastic body and partly as a rigid body. Figures 10–3 and 10–4 show the general features of earthflows. Material shears along the slip surface and slides as a more or less coherent body, perhaps itself cut by secondary slip surfaces. The downslope end of the earthflow spills over onto and slides along the old ground surface and is bent in the process, so that cracks and folds form as the result of tension and compression. Earthflows move in a range of velocities from a few feet a month to a few feet overnight, depending on the steepness of the slope and the properties of the material, which are primarily water-controlled. For example, a very soft, plastic clay soil may readily flow on any slope over 5°, whereas a less plastic, sandy soil may be stable on slopes of 15°. The size of the unit in motion ranges from about 4 feet wide and 6 feet long on new cuts with steep slopes to several

FIG. 10-4. View of cliffs formed at upper end of an earthflow as successive blocks moved downward on slip planes. General motion downward and to the right. (L. W. LeRoy.)

hundred yards wide and ½ mile long, measured in the downslope direction; a body of the latter dimensions might move a total of ½ mile under favorable conditions. Earthflows are often triggered by man-made excavations near the bottom of a slope. Movement may be halted by a number of devices (Fig. 10–5):

1. Reducing the weight of the moving body by removing water from it or removing material from the higher part.
2. Increasing resistance to gliding along the slip surface by drying it or adding additional load to the downflow part by piling more material on it.
3. Placing some kind of resisting structure at or near the toe; as, for example, erecting a retaining wall or driving piles through the earthflow into the underlying stable material.

In many situations where such measures succeed in stabilizing part of the earthflow, the rest may have been so weakened by its earlier movement that new slip surfaces readily form and the parts unaffected by the remedial action may continue to move.

Mudflows move much more rapidly than earthflows because they contain a higher percentage of water and thus have many of the properties of a liquid. The solid particles range in size from clay to large boulders. Soil under a sod or under a thin frozen layer may become saturated with water to the point of bursting through and flowing as a liquid down the surface until it reaches a slope so gentle that it stops (Fig. 10–6). Even

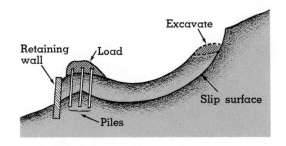

FIG. 10-5. Some methods of stabilizing an earthflow.

FIG. 10-6. Mudflows on nearly bare soil, frozen, then thawed by warm rain. (Soil Conservation Service.)

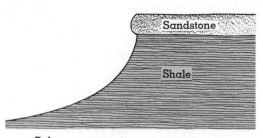

Before movement

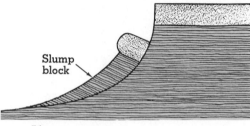

After movement

FIG. 10-7. Slump of horizontal sedimentary rock in cross section.

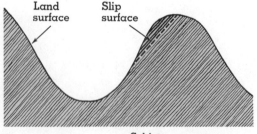

Schist
showing orientation
of cleavage

FIG. 10-8. Diagrammatic cross section of Hebgen Lake, Montana, rockslide. Material within dashed line slid into valley.

when containing boulders, a mudflow may be so fluid that it flows distances greater than a mile at speeds of 10 to 15 miles per hour. In Quebec, Canada, silt deposited on the bottoms of lakes long-since drained is a prominent source of mudflows. One is on record as having transported entire houses as far as 2000 feet within a few minutes.

Landslide is a general term covering the gravity transport of blocks of rock of various sizes and at various speeds. The category is divided into two subgroups separable on the basis of both of these criteria:

1. *Slump* is the comparatively slow movement of an unbroken block of moderate size along a curved slip surface with a substantial amount of rotation (Fig. 10-7). Movement normally occurs along joints, along bedding planes, or at the boundary between weathered and unweathered rock. Slump blocks range in size from a few feet to several hundred feet in their maximum dimension.

2. *Rockslides* consist of comparatively large blocks of rock moving as parts of a much larger mass, usually with extremely rapid motion. The slip planes are bedding planes, joints, or surfaces separating unlike materials. The Hebgen Lake, Montana, rockslide of 1959, set off by an earthquake, involved an estimated 80 million tons of weathered schist, which slid down the planes of schistosity into the adjacent valley, across the valley, and about 300 feet up the opposite slope (Fig. 10-8).

Occasionally one sees an example of mass movement that is interesting for the fact that it results from several changes in the environment, none of which is likely to have been effective if acting alone. The following is such an occurrence observed by the authors; it is illustrated in the map and cross section of Figs. 10-9 and 10-10, which show what appear to have been the significant relationships.

After a heavy late-spring snow which melted rapidly and infiltrated the gravel cap on the higher terrace, a series of springs appeared high on the upper slope. The soft shale was saturated with water, and the material indicated by diagonals slid off into the valley of Fountain Creek and took with it several hundred feet of the northbound lane of the highway. The combination of a gravel cap, narrow lower terrace, and an undercut bank of the valley occurred only at this point along the highway. Presumably the wetting and softening of the soft shale by meltwater, the undercutting of the bank at the outside of the bend of the creek, and the vibration of traffic, all acting in concert, produced the movement.

STREAMS

In man's early days his well-being was closely related in a variety of ways to the presence of streams. They were sources of fresh water and of food—mussels and fish

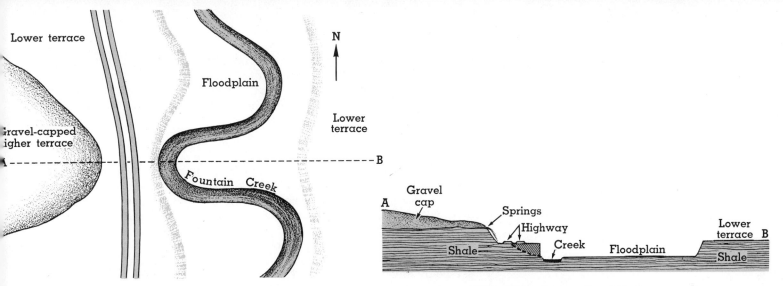

FIG. 10-9. Map of landslide area.

FIG. 10-10. Cross section A-B of landslide area, looking north.

as well as land animals who, like himself, came to the river to drink and on which he could prey. Flourishing agricultural civilizations developed in the fertile soils of river floodplains and deltas. The Nile, Tigris, and Euphrates rivers became part of the heritage of Western man both as centers of rich agricultural areas and as sites of great cities. Transportation too was concentrated along rivers, either in boats or by the use of pack animals and wheeled vehicles traversing the gently sloping, well-drained terraces of the river banks. Many topographic barriers otherwise impassable are breached by large rivers, and for centuries such gateways were vital to the exploration and settlement of new lands. Thus rivers have had a decisive influence on the commerce of the world before modern man developed means of travel not so closely tied to the characteristics of the earth's surface, and since then to a substantial if lesser extent.

DEFINITION AND GEOMETRY

A stream is any body of water, whether great or small, that flows overland. It is seldom constant in any property, changing direction, velocity, and form of channel all along its path. Streams flow in restricted channels, usually between comparatively high, steeply sloping banks.

Most streams are linear in form; that is, they have one long dimension and two short dimensions, like a pencil. Correspondingly, many landforms created by stream action are also linear: valleys, floodplains, terraces, and others. Because a river or a gully does not always stay in a direct path but migrates laterally, its effects may be discernible over a broad area. Further, since a major stream and its tributaries form an integrated network, the total area affected by the system is more extensive than the sum of its separate parts.

ENERGY AND MOTION

The force causing streams to flow is the gravitational attraction between the earth and the water on a sloping land surface. Gravity acts perpetually, without any change in its vertical direction. The steeper the slope the more nearly parallel the direction of flow to the direction of the force and the greater the velocity of flow, up to a point. Velocity cannot increase without limit because energy is lost by friction within the water and due to the roughness of the channel, so stream velocities seldom exceed 5 miles per hour. Smooth flow parallel to the channel surface is rarely seen. In most streams turbulent flow involving small and large eddies with horizontal, vertical, and inclined directions is the rule.

MATERIALS AND THEIR SIGNIFICANT PROPERTIES

In considering stream action, we can distinguish two categories of material: the fluid stream, which plays the active role, and the solid materials of the earth, which are acted upon.

Water at ordinary earth-surface conditions is a liquid; has a constant volume, within the range of normal temperature-induced expansion and contraction; and completely fills the lower part of its container, here the stream channel. Because it has mass, it has momentum when in motion, tends to remain in motion once moving, and resists changes in direction, as indicated by the comparatively rapid rate of erosion on river bends. Although it is lower in specific gravity than mineral fragments, it buoys them up significantly. Its viscosity is high enough that when in motion it can set fragmental material in motion, and at all times, whether moving or still, it offers resistance to solid particles falling through it. Pure water is a solvent of many minerals, and water when acid (either organic or carbonic) may be many times more effective in its dissolving power.

The solid materials acted upon by streams range in composition through the entire list of minerals and rocks; and in form from single molecules in solution through colloidal particles, the finest clastics, and boulders to jointed and unjointed bedrock in place. Particles that have been previously separated from the bedrock may be more-or-less readily picked up, transported, and deposited by water in motion, to an extent depending on the size, mineralogy, and shape of the particles and on the velocity of the water. Rock in place is primarily acted on by solution and abrasion, unless joints are so closely spaced that the blocks of rock react as fragments in an aggregate of particles. Rock properties that affect the extent to which solution and abrasion occur are hardness, solubility of the minerals, and kind and degree of cementation.

Sediment deposited by streams characteristically comprises rounded particles, sorted and stratified, in units of comparatively small size.

FUNDAMENTAL CONCEPT OF STREAM ACTION

A parachutist in free fall toward the earth reaches "terminal velocity"— his maximum rate of fall. In a similar way, particles of a particular size, shape, and composition have a characteristic *settling velocity*, which largely determines what happens to them when they are exposed to stream action. In turbulent streams, if the velocities of the upward currents in the many eddies exceed the settling velocities of available particles, the individual sediment grains will be picked up and transported. As long as the grains are passed from eddy to eddy, they will stay suspended in the water, settling to the bottom only when turbulence has decreased to the point that upward water velocities are less than particle settling velocities —or, eddies are so rare that they are inoperative in maintaining particles in suspension.

We can carry our analogy with the parachute jumper further: (1) The greater the specific gravity of the particle—equivalent to the jumper's weight as it might be augmented by strapped-on lead weights—the greater the settling velocity. (2) Just as opening the parachute vastly increases the jumper's area and thus his resistance to falling, variations in particle shape, and particle orientation if not equidimensional, also affect settling velocity. (3) The viscosity and specific gravity of the fluid are the final factors, as the parachutist would first be slowed and then buoyed up if he fell into water.

LANDFORM PATTERNS OF STREAM ORIGIN

The individual stream excavates its valley, which, like the stream, is basically linear in form. A stream drainage pattern is integrated in both the horizontal and the vertical; that is, in map view and as seen in cross section. Thus, in a river system one can go from any stream to any other without leaving the system, and all streams join the next larger or smaller stream at the same bed elevation (except for a few streams in hanging valleys, to be discussed in the next chapter). Despite many variations in stream patterns, the integration and completeness of drainage is always a prominent feature.

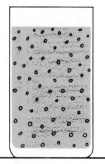

Sediment in
agitated water

Sediment deposited
in quiet water

FIG. 10-11. Effect of settling velocity on sedimentation.

The action of streams may be divided into two aspects: *Erosion* is the removal of sediment from its place of rest and its transport. *Deposition* is the laying down of sediment that has been transported.

An important feature of stream action is the process of *sorting*, in which particles of similar size, shape, and specific gravity are selectively eroded, transported, and deposited. Thus, for example, a volume of fragmental material composed of a wide range of different particles may separate into groups, with fragments of essentially the same settling velocity moving together—large, high specific gravity, equidimensional particles where velocity and turbulence are greatest, and small, low specific gravity particles of comparatively large cross section in relation to size where velocity and turbulence are least. Or, on the other hand, coarse gravel, fine gravel, sand, silt, and clay may all be well distributed from top to bottom of a stream when the water is strongly agitated, but when agitation stops the particles will settle to the bottom selectively—the coarsest first, then the next coarsest, and so on, the finest particles remaining suspended in the water for perhaps several days (Fig. 10–11).

Erosion. Erosion by streams includes several different, though often related, processes. The impact of raindrops may disturb particles and set them in motion. As rain falls, some water initially soaks into the soil (*infiltration*) and some flows over the surface (*runoff*) in a film of relatively uniform thickness. The layer of water may remove particles in a process called *sheet erosion*. Because this process lowers the surface uniformly its action is not very noticeable, but it is nonetheless effective. As the flow becomes concentrated in low places because of irregularities in the surface, rills are formed, and erosion proceeds there at an accelerated rate. Solid particles in transport may strike bedrock or soil and wear them away (*abrasion*). Another significant erosional process is *solution* of mineral material by water.

The turbulence of moving water is important in determining the extent of its eroding action. Where minor eddies are flowing faster than the main flow, particles can be picked up and transported that might otherwise not be disturbed. The size of available particles also controls the rate of erosion to some extent. Boulders cannot be picked up by currents that can rapidly erode sand; the same current may not be able to pick up clay because of its cohesive properties. At ordinary water velocities erosion of particles of intermediate size (0.06 to 0.5 millimeter in diameter) proceeds faster than erosion of either the finer or the coarser clastics. A vegetative cover reduces erosion by reducing impact and increasing infiltration.

Transportation. Moving water carries material from the place it was removed to the place it is deposited. Sediment is carried either in solution or as solid fragments. Most streams carry an appreciable quantity of mineral material in solution, and in clear streams this is the only manner of sediment transport. Calcite is the characteristic dissolved mineral of stream water, but often silica and occasionally a few other minerals are present in measurable amounts.

Clastic particles may be carried in three ways. (1) The finest particles remain in *suspension* by upward velocities of water that are greater than the settling velocities of the particles. (2) Larger particles may be lifted above the bed of the stream by locally stronger-than-average turbulence and travel horizontally while returning to the bed, where they may bounce back up or may dislodge another particle, which in turn is lifted and repeats the motion. This type of transport is called *saltation* (Fig. 10–12). (3) The largest fragments, which cannot be lifted by the turbulent water, slide or are rolled along the bed in *traction* (Fig. 10–12). The process of sediment transport is a flexible one—in constricted channels where velocities are high, saltation and suspension may be the dominant modes; if the channel is enlarged a few feet downstream, the same size particles may be carried by traction and saltation in the more slowly moving water. Stream-transported clastics are characteristically rounded by impact and abrasion.

The ability of a stream to move clastic sediment is described in two ways. (1) *Competence* is the diameter of the largest particle the stream can transport. Laboratory experiments and field studies indicate that with increasing water velocity, competence increases as the square of the velocity; i.e., if the velocity is doubled, competence increases by four, if the velocity is tripled, competence is nine times its former value, and so on. The competence of streams may be of surprising magnitude. Sediment samples taken from the Potomac River in flood stage have included boulders 15 inches in diameter that were netted 60 feet above the bed. Whether the boulders were in saltation or suspension, the turbulent energy available must have been very great. (2) The *capacity* of a stream is the total amount of sediment it can transport.

Both competence and capacity change as stream velocity changes; as a result, stream channels are different at flood times than in the low-flow stage (Fig. 10–13). When flow and velocity are low, sediments settle out and fill the channel. During floods, high competence and capacity result in a scoured-out channel.

The *longitudinal profile* of a stream—the outline of its channel showing the slope over a long distance—is almost always a curve that is concave upward, as in Fig. 10–14. The head of the stream is at A, its mouth at B.

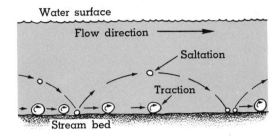

FIG. 10-12. Sediment transport in streams.

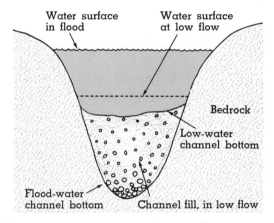

FIG. 10-13. Diagrammatic cross section showing differences in water surfaces and channel bottoms at flood water and at low water.

FIG. 10-14. Stream profile (vertical exaggerated).

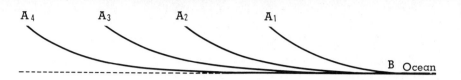

FIG. 10-15. Profiles of a stream at successive stages.

Many local variations exist but they do not substantially modify the overall shape of the curve. The *gradient* of a stream is the slope of the bed expressed in feet of altitude lost per mile of channel length. At the headwaters the gradient is steep, water velocity is high, and competence is high. Here the stream erodes rapidly, cutting its channel deeper and eroding it headward.

The curve of the longitudinal profile represents a dynamic equilibrium in which erosion and deposition are balanced in the short term and erosion then proceeds laterally. At this point the stream is said to be graded. Over the long range, the stream has the ability to downcut all or part of its channel. Consecutive positions of the profile of a stream flowing into the ocean illustrate the effects of downcutting (Fig. 10–15). Headward segments of the stream downcut rapidly, but note that the stream near its mouth does not. The profiles gradually approach the dashed line in the figure. For adjacent streams, the dashed lines together represent a smooth surface, gently sloping toward the ocean, called *base level.* Its special significance is that streams cannot cut below this level—it is the lowest surface to which the continents can be reduced by stream erosion. A similar relationship exists between the streams flowing into a large lake and the water level in the lake. This establishes what is called a *local base level,* or *temporary base level,* which may change as the level of the lake surface changes. In addition, the level of the main stream controls the profile of tributaries entering it, each major tributary controls the profiles of its own tributaries, and so on, down to the smallest rill. Thus we have an integrated stream system in which every element characteristically enters its master stream at the elevation of the master stream at the junction.

Valley formation. Active erosion by a stream above its base level is almost entirely vertically downward, deepening the valley in which the stream flows. The valley is also widened, but this is done for the most part by mass movement, small tributaries, and sheetwash. It is thought that the growth of a stream valley follows the sequence shown in Fig. 10–16. Initially the stream flows in a narrow, shallow, steep-walled valley, far above base level. After the graded profile is achieved, lateral erosion can begin. The stream downcuts, while maintaining constant side slopes, until base level is reached. At this point downcutting ends and erosion takes the form of sidecutting. The path changes from relatively straight to curved, and the stream undercuts the valley slopes, which are further eroded by mass movement; the valley walls retreat, maintaining a constant slope;

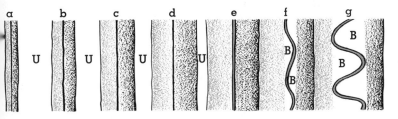

FIG. 10-16. Diagrammatic profiles across a stream valley.

FIG. 10-17. Map of stream valleys.

East-facing valley wall   U   Original plain

West-facing valley wall   B   Floodplain at base level

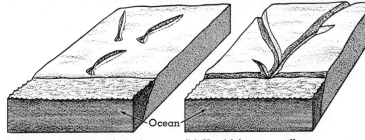

(a) Plain recently uplifted above base level, erosion beginning; short V-shaped valleys cut where runoff is concentrated in rills.

(b) Youthful stage, valleys connected, downcutting dominant in valleys.

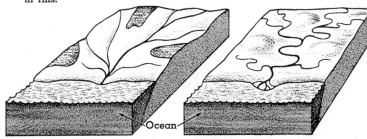

(c) Mature stage, sidecutting beginning.

(d) Old-age stage, base level reached in most of area, plain regained.

FIG. 10-18. Stages in the cycle of erosion in a humid climate.

and the river swings back and forth, creating a broad, flat *floodplain* at the altitude of base level. As time goes on, essentially all the rock and soil above base level are removed, and the river then winds sluggishly back and forth in great sweeping curves at base level, no longer eroding as its course changes except for local cut-and-fill action. An alternate hypothesis of slope retreat holds that slopes of valley walls decrease as time goes on. Slope appears to be controlled by a combination of factors, including weathering rate, rock type and structure, and climate. In plan view (as on a map) the stream paths and valley dimensions corresponding to Fig. 10–16 change as shown in Fig. 10–17.

With sufficient passage of time the entire area may theoretically be reduced to base level. If the surface is large, it is called a *peneplain*, meaning almost a plain (from the Latin *paene*, almost). Vast areas surrounding Hudson's Bay, Canada, are so near base level that this region may be a peneplain. We find isolated fragments of such surfaces in several areas of the world, and each has been a source of controversy as to the validity of applying the term.

The cycle of erosion. Laboratory studies of stream erosion and field observations of streams have made inescapable the hypothesis of gradual evolution of both streams and the topographic forms they carve. Figure 10–18 shows diagrammatically the stages of topographic development of a

TABLE 10–1   PRINCIPAL CHARACTERISTICS OF LANDSCAPE
IN THE STREAM CYCLE OF EROSION
IN A HUMID CLIMATE

FEATURES	YOUTH	MATURITY	OLD AGE
Stream courses	Few, short, straight	Many, some meandering	Fewer than in maturity, meanders numerous
Gradients	Steep, many falls and rapids	Intermediate	Gentle
Valleys	V-shaped (Fig. 10–19)	Appreciable floodplain widths	Very broad, floodplains of great width
Slopes	Steep, short, a small percentage of total area	Nearly all of area in slope, drainage completely integrated	Gentle slopes, a small percentage of total area
Divides	Broad, flat, a large percentage of total area	Occasional patches and strips of original plain on divides, most sharp	Broad, gentle, difficult to find position
Relief	Low at beginning, may become large	Maximum	Low

stream system in a humid climate (where precipitation exceeds evaporation and year-round streamflow is maintained). The hypothetical cycle begins with a nearly flat plain that is elevated above base level, so that erosion can occur. At the end of the cycle the area has returned to a peneplain, essentially at base level with occasional hills called *monadnocks*. The development is divided into stages called *youth, maturity*, and *old age*. Table 10–1 lists the principal features characteristic of each stage. As we learn more about stream erosion we realize that the process has complexities not encompassed in our present hypothesis. Nevertheless, the three-stage concept is a useful approach to the description of stream development.

In some areas we find evidence that the cycle of erosion has been reactivated after having once run its course. For example, an old-age stream with many meanders may exist in an area that is suddenly uplifted above base level. The stream immediately reacts by downcutting and thus

FIG. 10-19. V-shaped youthful stream valley. Lodore Canyon of Green River, Utah. (The American Museum of Natural History, New York.)

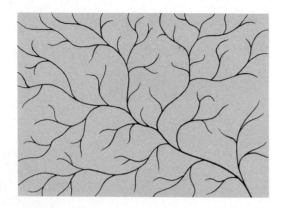

FIG. 10-20. Stream pattern on massive igneous rock and horizontal bedded rocks.

FIG. 10-21. Cross section of an unsymmetrical ridge formed on gently dipping bedded rocks.

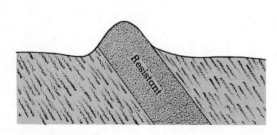

FIG. 10-22. Cross section of a symmetrical ridge formed on steeply dipping bedded rocks.

erodes a narrow, V-shaped valley that is typical of youth as seen in cross section but preserves the meander pattern of old age as seen in map view. This anomalous combination characterizes a stream that is said to have been *rejuvenated*. The same result can be produced by the sudden lowering of a local base level, as where a lake has been drained rapidly and streams flowing into it are rejuvenated.

Stream patterns. The materials over which streams flow cover a wide range: thin soil; thick soil; batholiths; horizontally bedded, folded, or faulted sedimentary rock; and many others, each material and structure having its effect on the landforms and drainage network that develop on it. Batholiths, horizontally bedded sediments, and extrusive igneous rocks are isotropic in the horizontal direction, so the stream patterns they influence have no preferred direction. Figure 10–20 shows the uniform density and spacing and the random direction of such streams. Where sediments are tilted, different patterns result. In Fig. 10–21 we see the topographic profile and cross section characteristic of gently dipping beds that are alternately resistant and less resistant to erosion. The resistant rocks stand up in higher relief, forming ridges with a short, steep slope on one side and a long, gentle slope on the other, in the direction the beds are dipping. Such an unsymmetrical ridge is found where dips are less than 30°. Where dips exceed 45°, the ridges are more nearly symmetrical (Fig. 10–22).

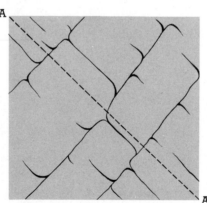

FIG. 10-23. Drainage pattern on dipping bedded rocks.

FIG. 10-24. Topographic profile of Fig. 10-23 along line A-A'.

FIG. 10-25. Cross section of Fig. 10-23 along line A-A'.

FIG. 10-26. Ridges on resistant rock and valleys in less resistant beds dipping to right. (The American Museum of Natural History, New York.)

If the strike of the beds is constant except for small local variations, the main streams are subparallel and tributaries enter at about right angles, as in the drainage pattern shown in Fig. 10–23. The lengths of the tributaries indicate that ridge crests are not midway between main streams, and the topographic profile (Fig. 10–24) shows unsymmetrical ridges. Filling in the strata below the surface shows an anticline (Figs. 10–25 and 10–26). Similarly, stream patterns may be used to recognize synclines or dips in uniform directions. If the structure is a dome or basin, ridges are roughly circular or elliptical, tracing the form of the fold, and concentric stream patterns result (Figs. 10–27 and 10–28).

If one or more sets of joints occur in rocks in which a dendritic drainage pattern would normally develop, the joints tend to control the orienta-

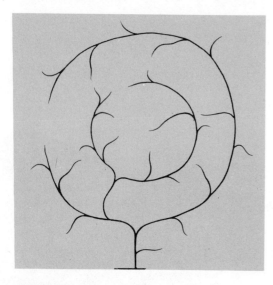

FIG. 10-27. Concentric drainage pattern on a dome.

tion of valleys and to establish an angular pattern (Fig. 10–29). Numerous parallel straight channel segments exist where joints have controlled weathering and erosion.

Deposition. In the early stages of stream development, and in the up-stream sections of streams, erosion is the dominant process, but any stream carrying sediment can deposit it if conditions are favorable. As gradients lessen and streams begin to erode laterally, deposition of sediment becomes more and more important, and this continues as the system ages. Because deposition occurs rapidly, our understanding of it is better than of the erosional process.

Sediment may be deposited to form an obstruction in the channel at the site of a local decrease in velocity that has caused reduction of competence and capacity. The obstruction tends to grow by continued deposition in the sheltered area on its downstream side. The deflected currents impinge upon the sides of the channel and erode material, which is transported downstream from the site of erosion. It may be deposited in quiet water just downstream from the area it came from; the current there is then deflected toward the opposite side of the channel, and the process is repeated. The sequence of development of stream curves is illustrated in Fig. 10–30. The channel cross section at a bend is shown in Fig. 10–31. Because of centrifugal force, maximum velocities and turbulence occur in the deepest part of the channel near the outside of the bend. Erosion

FIG. 10-28. Subcircular ridges on resistant sandstone and concentric drainage on eroded anticline. Middle Dome, near Harlowton, Montana. (The American Museum of Natural History, New York.)

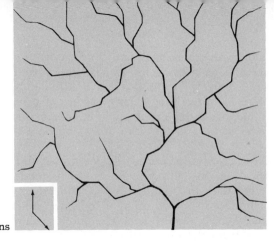

Joint
directions

FIG. 10-29. Angular drainage pattern.

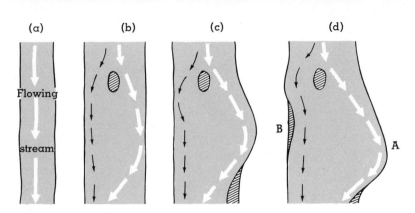

FIG. 10-30. Development of curves in a stream. Narrow arrows: weak current; broad arrows: strong current; lined areas: sediment deposition.

dominates there, the bank is undercut at point A [Figs. 10–30(d) and 10–31], and the material removed is deposited in the low-velocity water on the inside of the next bend downstream (point B). The high-velocity thread of water crosses the channel to the outside of the next bend. Bends continue to be widened, forming the broad, sweeping curves called *meanders*, and move downstream (Figs. 10–32 and 10–33). This estab-

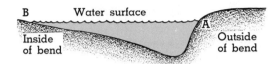

FIG. 10-31. Cross section of stream at a bend.

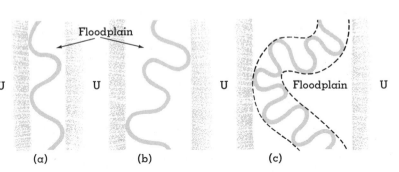

FIG. 10-32. Maps showing growth of a meander belt. Lined areas: sloping valley walls; U: flat-topped divides; meander belt, inside dashed lines.

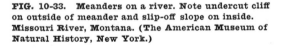

FIG. 10-33. Meanders on a river. Note undercut cliff on outside of meander and slip-off slope on inside. Missouri River, Montana. (The American Museum of Natural History, New York.)

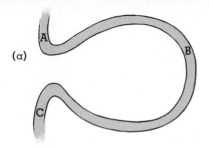

(a)

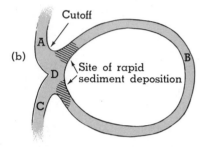

(b)

Cutoff

Site of rapid
sediment deposition

FIG. 10-34. Cutoff and oxbow lake.

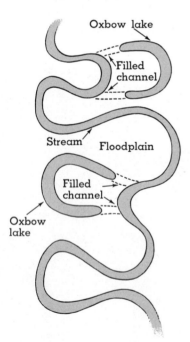

Oxbow lake

Filled
channel

Stream Floodplain

Filled
channel

Oxbow
lake

FIG. 10-35. Meandering stream and oxbow lakes.

lishes a *meander belt*, which slowly migrates back and forth across the floodplain, and widens wherever erosion acts on the sloping valley wall. The site of erosion is not controlled solely by the position of the bends, for in some areas the material is the decisive factor. Large bodies of compacted clay in the floodplain sediments are difficult to erode in comparison with more easily removed silt and sand, and they can act as barriers to the migration of the channel.

Easily erodible sediment in the narrow strip between adjacent meanders may be removed to form a *cutoff* (Fig. 10-34). Once the breach is made, the gradient A-D-C is much steeper than the gradient A-B-C, and velocity is hence much greater in the cutoff than in the original meander. Sediment is deposited in the meander, most rapidly near the cutoff, where clastic particles constantly escape from the rapid current into the nearly still water of the meander; the channel becomes filled in at this point, and eventually an *oxbow lake* is formed in the abandoned meander (Figs. 10-35 and 10-36).

During floods the river may overflow its channel and spread across the floodplain. Water velocity is high in the channel but decreases sharply over the vastly larger area of the floodplain. Sediment is deposited as a result of decreased velocity and competence, producing a wedge-shaped deposit, thicker and coarser-grained near the channel and thinner and finer-grained toward the valley wall (Fig. 10-37). The surface is highest near the channel, and as the channel flow returns to its original cross section, sediment may be deposited in the channel as well as on the floodplain. The raised strip is called a *natural levee*, and it is commonly found where stream deposition dominates over erosion. If natural levees are large, the stream water surface may be several feet higher than the land surface on the distant edge of the floodplain.

Natural levees are desirable sites for human habitation, as the surface is high and dry, it is better drained because of the coarser grain size of sediment and its sloping surface, and the coarser sediment makes a better foundation material on which to build than the clayey and organic sediment in the adjacent floodplain. Buildings and roads are restricted to the natural levee if the adjacent floodplain is swampy; but if the floodplain is adequately drained, it is favorable for agriculture and habitations, even in the face of occasional floods, because weathering of the mineral mixture in the floodplain sediment provides a great number of nutrients.

Deltas are depositional features formed where streams carrying clastic sediment empty into the comparatively quiet water of lakes or oceans. Where the water velocity lessens, competence and capacity decrease and grains settle to the bottom. An additional factor in construction of a delta along the ocean shore is the clumping of clay particles in salty water and their more rapid settling out than when dispersed.

FIG. 10-36. Oxbow lake nearly completely filled by sediment. (U.S. Department of Agriculture.)

Growth of a delta takes place as illustrated in Fig. 10–38. The fine cross-hatching represents sediment deposited as a delta. In stage c, the gradient from point P to the end of the delta is less than the gradient from P to Q. During a flood the overflow may take the path P to Q and because of the steep gradient will rapidly erode a new channel. This will become the dominant one owing to its steeper gradient (stage d). In another flood the process is repeated and a new channel is formed on the line P-R [Fig.

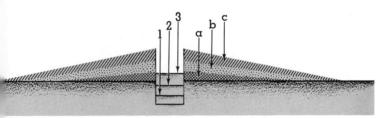

FIG. 10-37. Diagrammatic cross section of natural levee. a, b, and c are consecutive layers deposited (vertical exaggerated). 1, 2, and 3 identify channel bottoms corresponding to natural levee deposits.

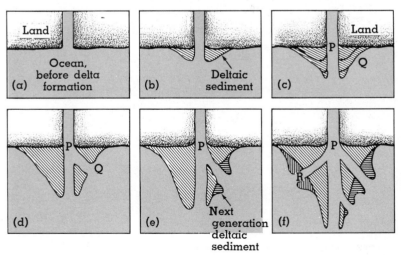

(a) Ocean, before delta formation
(b) Deltaic sediment
(c) Land P Q
(d) P Q
(e) Next generation deltaic sediment
(f) R P

FIG. 10-38. Growth of a birdfoot delta, map views.

FIG. 10-39.
Map of Nile delta.

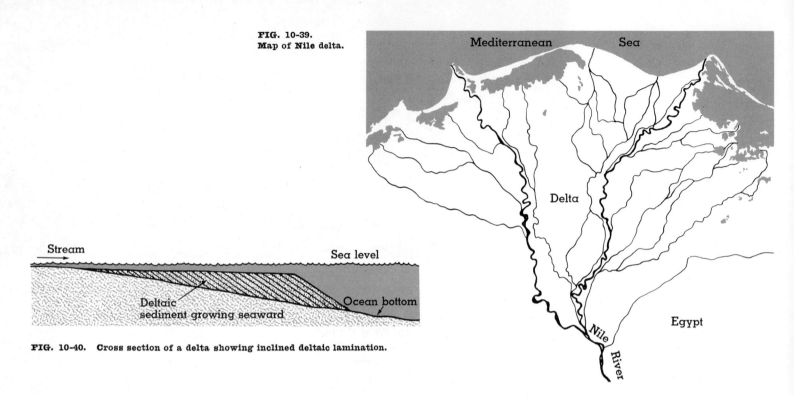

Stream

Sea level

Ocean bottom

Deltaic sediment growing seaward

FIG. 10-40. Cross section of a delta showing inclined deltaic lamination.

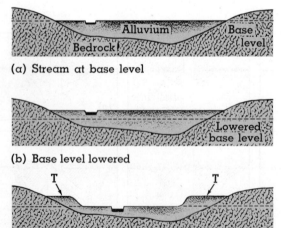

(a) Stream at base level

Alluvium

Bedrock

Base level

(b) Base level lowered

Lowered base level

(c) New floodplain at lower base level

T   T

FIG. 10-41. Stream terrace formation.

10–38(f)], where the gradient is steeper than in the existing channels. The numerous channels through which the water flows into the larger body of standing are called *distributaries*. A delta of this kind, with linear deposits of sediment along distributaries, is called a *birdfoot delta* from its resemblance to the clawed bird foot. It forms where the deltaic deposits are not shifted by wave action. The delta of the Mississippi River is an excellent example. The name delta (from the Greek letter Δ) was derived from the triangular deposit formed by the Nile River (Fig. 10–39). A cross section of a delta (Fig. 10–40) shows the lamination of sediment to be more steeply dipping than in the sediment deposited on the adjacent bottom of the standing water body.

If the delta surface is high enough so that floods and tides seldom disturb man's use of it, the combination of a heterogeneous, fine-grained parent material, occasional replenishment of material by floods, available fresh water, a flat surface, harbor facilities, and convenient communication routes both landward and seaward combine to make the delta desirable for human occupancy. Evidence for this exists in the long-inhabited densely populated deltas of the Waal and Rhine in Europe, the Indus and Ganges of India, the Hwang Ho and Yangtze of China, and the Nile of Africa, and the recent struggles for possession and control of the delta of the Mekong River in Vietnam.

The cross section of many rivers in the old age stage is as illustrated in Fig. 10–41, which shows a large valley eroded in bedrock on which there is a layer of *alluvium* (stream-deposited sediment; see Fig. 10–42) and a river flowing in a channel. If the floodplain is essentially at base level, the stream is sidecutting and not downcutting. If base level is lowered by a substantial amount (or the area raised), as indicated in Fig. 10–41(b), the stream will commence downcutting and will continue until it reaches the new base level, when it will return to sidecutting. Figure 10–41(c) represents landforms associated with the stream after it has formed an extensive floodplain at the new base level. Remnants of the earlier floodplain exist at T and T′ as paired terraces. Depending on the sequence of events, remnants may, at any given time, exist on both sides of the valley, on only one side of the valley, or not at all—even though they are present

FIG. 10-42. Alluvium deposited on a preexisting floodplain, Wisconsin. Upper layer built up in 70 years. (H. H. Bennett.)

at other places up or downstream. Repeated intermittent uplifts of the land (or lowering of base level) result in the formation of corresponding numbers of terraces; in some areas as many as seven sets of paired terraces may be seen in river valleys (Fig. 10–43). Terraces have many features that make them favorable for human habitation: The surface is safely above all but the highest of floods; the veneer of sediment is well drained and is therefore agriculturally desirable and also forms stable foundation material; soils are usually fertile because of the heterogeneous composition of alluvium; sand and gravel deposits provide raw materials for construction; the surfaces are extensive and nearly horizontal, permitting efficient use of farm machinery and providing the best sites for highways, railroads, canals, pipelines, and transmission lines; and the river may also be a communication route.

Youthful rivers flowing in mountain ranges have comparatively steep gradients and therefore have a high competence, even though their capacities may be much less than those of larger streams of lower velocity. Where mountain streams flow out on the gently sloping plains adjacent to the ranges, their velocities decrease and the coarser sediment is deposited in a conical pile. This accumulation of sediment is called an *alluvial fan* (Fig. 10–44). It occurs roughly as indicated in the cross section of Fig. 10–45, in which the vertical dimension is greatly exaggerated. Alluvial fans are built up much like deltas: by streams depositing sediment along their courses, repeatedly shifting to another course and depositing adjacent to it, shifting again, and so on. Stream deposits, like streams, are linear, so although the cross section of the alluvial fan shows layering inclined from the apex down to the toe of the fan, extensive beds do not exist. As one might expect, the coarsest sediment is deposited first and occurs at the apex of the fan, and sediment is finest at the toe of the slope. In most places alluvial fans are built up rapidly during flood times and slowly when stream flows are low. In times of low flow, streams may even sink into the coarse sediment near the apex of the fan, and only at times of maximum runoff will they flow on the surface to the toe of the fan. Fans side by side along a mountain front may merge to form a continuous slope.

Alluvial fans range in size from a few feet from apex to toe to the tremendous fans in the Great Valley of California and the arid basins of the Middle East, where the apex may be 2000 feet higher than the toe and 20 miles from it. In such arid and semiarid places the alluvial fans are agriculturally the most productive landforms because of the easily tilled slope (more gentle than in the mountains); the rich soil formed on the lower slopes from the fine, heterogeneous sediment; the presence of fresh water originating in the mountains; and a surface that is high enough to be away from the saline lakes or soils found in the centers of many arid valleys.

FIG. 10-43. Stream terraces, Snake River, Grand Teton National Park. (National Park Service.)

High-velocity mountain stream

Coarsest sediment

Medium-sized sediment

Finest sediment

Bedrock

FIG. 10-45. Cross section of an alluvial fan.

SUBSURFACE WATER

Subsurface water is the water that exists in openings in soil and rock below the land surface. Its prime significance in our everyday lives is as a water supply—in many areas the only one it is practical to use. The presence of subsurface water is of concern to the engineer, who must consider its effects on foundation materials, and is also of importance for the accumulation of some minerals of outstanding economic value; for example, copper in the southwestern United States. Its capabilities as an agent responsible for landform development are most strikingly exemplified by such great limestone caverns as the Carlsbad Caverns of New Mexico and Mammoth Cave in Kentucky.

Ground water is that part of the subsurface water that is in the saturated soil and rock, beneath an upper zone through which water percolates without being held.

GEOMETRY

The shape of the subsurface water body is, on a fine scale, that of the openings which contain it—the spaces between grains in a sandstone or in joints, or dissolved openings in soluble rock. On a larger scale the subsurface water occupies a complex three-dimensional network consisting of the connected void spaces in soil and rock, limited above by the land surface and below by the absence of open spaces.

ENERGY AND MOTION

The force of gravity causes the motion of subsurface water. Thus, when free to move in rock openings it seeks the lowest level. Because most openings are small, its velocity is usually low, perhaps a few inches or a few feet per year, but in large cavities such as caves it may move at rates comparable to the velocities of streams. Ultimately, as it continues to move in a downslope direction, it will reach the sea; but its path may be extremely tortuous, depending on the available openings, in much the way water flowing through a coiled hose many feet long must follow all the bends and coils before reaching a nozzle that may be only a few feet from the faucet. In contrast to streams, subsurface water may have great freedom of action in the vertical direction.

MATERIALS AND THEIR SIGNIFICANT PROPERTIES

Water, materials dissolved in it, and the solid materials of soil and rock it encounters are the substances involved in subsurface processes. Water dissolves minerals, transports them in solution, and deposits them. The greater the amount of carbon dioxide and organic acids it contains, the greater its ability to erode by solution; and where the soil and rock are made up of the more soluble minerals, the processes of solution will be most conspicuous. Because, as you will recall (Chapter 9), solution depends on the existence of substantial contact area, bedding and jointing of rock are of great significance, and thinly bedded or closely jointed rock is most susceptible to solution. Clastic sediment is almost never transported in this environment because water velocities are too low.

FUNDAMENTAL CONCEPTS

Because water moves so slowly underground, solution is the significant action there. It goes on through three dimensions, and acts equally rapidly and to the same extent in all directions. The process has the greatest effect where the most soluble minerals exist and where bedding and jointing of rock permit the greatest penetration of water.

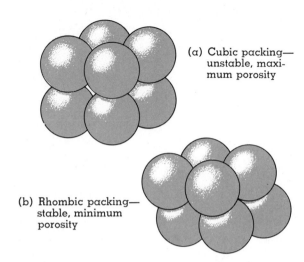

(a) Cubic packing—unstable, maximum porosity

(b) Rhombic packing—stable, minimum porosity

FIG. 10-46. Packings of spheres of uniform diameter.

Landform patterns take a random form except that solution features are limited to regions where soluble rock occurs and may be more numerous where water is concentrated, as in joints. Surface drainage, if present at all, is characterized more by nonintegration than by integration, and much water is likely to go underground.

Two properties of solids are of first importance in discussing the action of subsurface water: porosity and permeability. *Porosity* is the property of containing voids and is stated as a percentage of the total volume of material. In igneous rocks, original porosity exists in the spaces between mineral grains (usually of microscopic size and less than ½ of 1 per cent of the rock in volume) and in the vesicles of scoria and pumice. In clastic sediments, clastic sedimentary rocks, and pyroclastics the original voids are the interstices between the particles. The degree of original porosity that may occur in clastic sediments covers a wide range. Let us imagine a sediment composed of spheres that are all the same size. If the spheres were to be arranged, or *packed*, so that their centers occupied the corners of cubes, as shown in Fig. 10–46(a), the porosity would be about 48 per cent; but this packing is not stable. Centers of spheres are above centers, and if there is some disturbance, each sphere will tend to slip down to fill the space between spheres in the layer beneath; thus each layer will be repacked in a more compact pattern, as in Fig. 10–46(b). In the second arrangement the porosity is close to 27 per cent.

This range of porosity depends on packing alone; that is, with spheres of uniform size in an aggregate, porosity is determined only by packing. Hence a tank full of spherical lead shot will have the same porosity as a tank full of basketballs if the packing arrangements are identical. Of course, if other spheres of smaller size are in the spaces between the largest spheres, and even smaller spheres fill the smaller voids, and so on through successively smaller diameters, the porosity of the aggregate will be far less than if only one size of sphere is present. Further, the shape of the particles affects porosity. Perfect cubes of uniform size can be packed so that the porosity is essentially zero, whereas platy grains will give porosities of 40 to 75 per cent (Fig. 10–47). Clay particles, which are platy, are so small that surface forces are significant in maintaining their arrangement. A fresh deposit of clay can have such an open-work, chain-type structure that it will have a porosity of over 90 per cent (Fig. 10–48). This high porosity may be reduced if the particles are disturbed and repacked, or may even be increased by repacking. Earthquake action and mass movement often have such effects on uncemented sediment. Naturally occurring clastic sediments seldom consist of spheres of the same diameter; and their porosities are seldom more than 25 per cent and most often less than 15 per cent. In most metamorphic rocks, original porosity is extremely low, owing to the loss of pore space by pressure or addition of matter in their environment of formation.

Secondary porosity is not uncommon and is more significant than original porosity in most igneous and metamorphic rocks. You will recall that brittle rocks of all kinds may rupture under folding or compression to form joints (Chapter 8). If the joints are numerous and closely spaced, they may make up nearly all the porosity of a rock, whether igneous, sedimentary, or metamorphic. Basaltic lava flows characteristically are prismatically jointed because of tension cracks that form in a honeycomb pattern when the flow cools after solidifying; this is a widespread type of porosity in basalt. Soluble rocks such as limestone may contain many openings formed by solution. Such cavities range from microscopic size to chambers of the dimensions of the Big Room in Carlsbad Caverns, which is nearly 4000 feet long, 600 feet wide, and 300 feet high. Porosity can also be reduced by secondary processes such as deposition of a mineral that may fill pores or may cement a clastic sediment—for example, sandstone, which seldom has a porosity of more than 22 per cent. Porosity of any origin provides storage space for water. The connecting openings between voids may be small, but in the great length of geologic time even slow movement of water can be effective in enlarging the pores by solution or reducing them by deposition.

Permeability is the capacity of materials to transmit fluids. A material of high permeability transmits water rapidly, one with low permeability transmits it slowly. The size of the openings is important in controlling permeability, for water will not move if the voids are too small. Thus, although porosity is not related to particle size, permeability is. Resistance to flow is caused by the friction of fluid against the surfaces of the conduits. Coarse-grained sediments such as gravel, which have large intergranular openings, are therefore more permeable than fine-grained sediments such as silt. Doubling the particle size will increase permeability roughly about seven times, other things being equal. If there is to be a large permeability, even in a very porous material, the voids must be connected; and for greatest efficiency, the connections must be numerous and comparatively large in diameter.

A water well is drilled until it reaches a layer of material that will yield water. A stratum that produces water in economic amounts is known as an *aquifer*. A gravel of high permeability may be an aquifer even if it is only a few feet thick, whereas a medium-grained cemented sandstone may have to be many feet thick in order to produce water in comparable amounts. In igneous rocks such as granites and metamorphic rocks such as gneiss or quartzite, all the porosity and permeability may be concentrated in widely spaced fractures. A well may penetrate hundreds of feet

FIG. 10-47. **High porosity of platy and linear clastics, shown in cross section.**

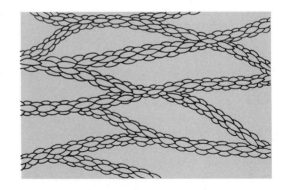

FIG. 10-48. **Extremely high porosity of flocculated clay, shown in cross section. Each line when magnified has the structure of Fig. 10-47.**

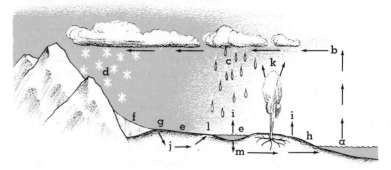

FIG. 10-49. Major components of the hydrologic cycle, diagrammatic and much simplified.

of dry rock in which porosity and permeability are effectively zero before reaching one fracture which, acting as a conduit, may be a very productive source of water. If rock is broken and shattered in a fault zone, the zone may be a conduit of high permeability, but if the shattered rock has been crushed to a fine powder or altered to a clayey gouge, it may act as a low-permeability barrier to water flow.

THE HYDROLOGIC CYCLE

Subsurface water is not formed in place, nor does it simply occur like blood in veins in the earth, as has been at times suggested. Beginning with the early Greek scholars, its origin has been a subject for study. We now know that most subsurface water is involved in a great cycle of moving water, as illustrated in Fig. 10–49:

Water is evaporated from great areas of the ocean surface (a), changing from the liquid state to vapor, and some is transported by winds (b) to the continents. There some falls as rain (c), some as snow (d), and some as other forms of precipitation. Some of the rain strikes the land surface and is absorbed, and some is caught in lakes (e) and streams (e), joining those bodies of surface water. Snow may accumulate in snowfields (f) and glaciers and be stored there for many years. Some snow melts, part to run off as surface water (g) and (h) to the sea by gravity flow. Some water is evaporated (i) en route and joins the air-circulated water vapor, to be recycled. Ground water, the water from rain, snowmelt, or surface water that saturates soil and bedrock, is also in motion. Acted on by gravity, it moves downward and laterally (j) toward the oceans. At times there may also be upward motion. Some of the water absorbed by the soil is returned to the air by evaporation. Plants withdraw some water and transfer (k) it to the air as vapor. There is some interchange between surface and subsurface water, as by spring flow (l) or infiltration (m) from lakes and streams.

266

We have here identified the most significant parts of the *hydrologic cycle*, the dynamic equilibrium participated in by the water of the earth. Other paths that water may take have been omitted for the sake of simplicity.

DISTRIBUTION AND OCCURRENCE OF SUBSURFACE WATER

We find that the distribution of subsurface water falls into zones, each with its own characteristics. Figure 10–50 shows these zones diagrammatically. It is not drawn to scale because the zones are of irregular thickness, and some may be absent altogether, depending on local conditions.

If in a small area a group of wells is drilled to sufficient depth, they will probably all reach water. The three-dimensional surface connecting the water levels in the wells is called the *water table*. It is the dividing surface between the perpetually saturated materials below and the usually unsaturated materials above. The region above the water table is called the *zone of aeration*, because the pore spaces are usually partly or entirely filled with air. The zone below the water table is the *zone of saturation*; here the connected pore spaces are entirely filled with water—the ground water. The materials at and near the surface support vegetation. In this *soil zone* the environment changes rapidly. During and after a rain or when snow is melting, water infiltrates the soil and drives out the gases produced by chemical processes or by life forms. Evaporation and plants may withdraw the soil moisture, and the voids are then filled with gases again. Some moisture moves downward beyond the reach of the plants,

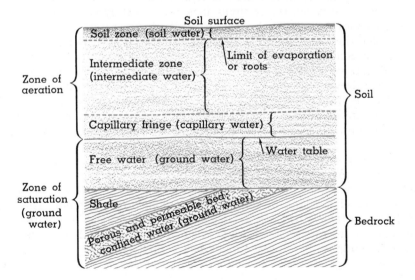

FIG. 10-50. Diagrammatic cross section showing occurrence of subsurface water.

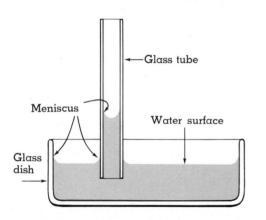

FIG. 10-51.   Capillary rise of water in a glass tube.

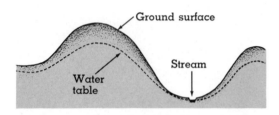

FIG. 10-52.   Cross section showing water table.

taking with it dissolved materials. Thus, in this active zone frequent change is the rule.

Just above the water table, water is lifted upward by *capillary action*. There is an attraction between the molecules of water and the molecules of the minerals great enough to overcome the downward pull of gravity. Figure 10–51 illustrates, with some vertical exaggeration, this effect. A slender glass tube is in water in a glass dish. Around the edge of the dish and the outside of the tube, water is drawn upward by cohesion and adhesion to form a meniscus. Because there is little water inside the tube compared with the total area of water in contact with glass, water is lifted to a higher level on the inside than on the outside, where there is much more water in comparison with the total contact area. The smaller the diameter of the tube the further water can be raised.

In a similar way, the interstices in clastic sediments make up a three-dimensional network of tubes, in which water is raised above the water table. Hence this zone is called the *capillary fringe*. For any specific particle size and diameter of opening there is a maximum distance water can be raised by capillary action. The combination of grains and openings of many different sizes causes water to be raised more in some openings than in others; hence the amount of water in the capillary fringe will range from near-saturation just above the water table to a comparatively low percentage some distance above it. The thickness of the capillary fringe ranges from a small fraction of an inch in coarse gravels to over 60 feet in silts and clays. The water table may rise when the supply of water is abundant and drop when large amounts of water are pumped from wells. The capillary fringe rises and falls with the water table, so this zone is fairly active; at times it is nearly saturated with water and at other times contains soil gases in the openings between the grains.

Between the capillary fringe and the soil zone is the *intermediate zone*, a body of material that is essentially unchanging in its properties. In some situations it may be absent, as where the water table is 15 feet below the ground level and the soil is so fine grained that the capillary fringe would be 16 feet thick. Under such circumstances, and as long as temperatures are above freezing, there will be a constant movement of water vertically upward to the surface, where it will evaporate or will be used by plants.

The water table is a three-dimensional surface that may be contoured like the ground surface. It is generally high where the ground surface is high and low where the ground surface is low (Fig. 10–52), resembling the surface of the land in general form but having less relief. The slope of the water-table surface is related both to the permeability of the sediment and to the rate at which water is moving through it. If the water-table slope is gentle, either the sediment has a high permeability or water is flowing through it at a low rate. A steep water-table slope suggests the reverse: a low permeability or a high rate of flow of water through it.

The water table may intersect the land surface. A body of surface water is an example of such an intersection. If the water table around the body of water slopes downward toward it, ground water flows down the water-table slope, comes to the surface, and maintains the surface-water body, whether a spring, lake, or stream [Fig. 10–53(a)]. This is called an *effluent* condition and is characteristic of humid regions. If the water table slopes downward away from the body of surface water, the lake or stream loses water by infiltration into the surrounding sediment [Fig. 10–53(b)]. This *influent* condition is typical of arid regions.

The water in the zone of saturation occurs in two very different sets of conditions. Water below the water table and above the bedrock surface moves downward and laterally under the force of gravity. It is called *free water* because its motion is comparatively unrestricted. If you examine the bedrock of Fig. 10–50, you will note that a porous and permeable bed lies between two impermeable beds and is also in contact with the over-lying saturated, uncemented soil. Free water can move downward from the soil into the porous and permeable bed. But it is then restricted in its flow to the thickness and areal extent of the bed. Thus it is called *confined water*, and it flows under pressure through the bed just as water flows through a pipe—sometimes down, sometimes up, sometimes horizontally, depending on the attitude of the bed. This condition is basic to the operation of the artesian well, as indicated in Fig. 10–54. The cross section

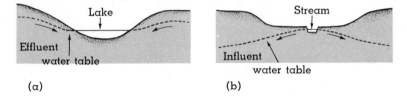

(a)   (b)

FIG. 10-53. Effluent and influent water tables. Arrows show direction and flow of ground water.

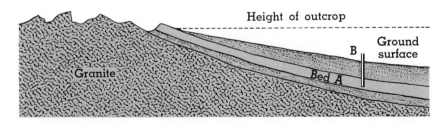

FIG. 10-54. Cross section of artesian basin.

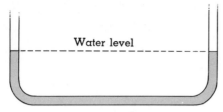

FIG. 10-55. Water in a U-tube.

shows a mountainous area with a granitic core, surrounded by ridges on resistant sedimentary rocks and a plain sloping gently downward away from the mountains. Rainwater, snowmelt, and influent streams contribute water to bed A where it is exposed, and the confined water in it flows downdip. A well drilled at point B penetrates the aquifer bed A. The water, under pressure, will theoretically rise to the height of the outcrop of bed A, just as water in the two ends of a U-tube will be at the same height (Fig. 10–55). This may not actually occur, because of frictional losses, but the water will at any rate rise above the level of the aquifer at the point of penetration. This geologic structure and environment is called an *artesian basin*. In technical usage, we speak of a *flowing well* if the water rises to the land surface, and an *artesian well* if it rises above the top of the aquifer but not to the land surface.

LANDFORMS CAUSED BY GROUND WATER

Ground water, although out of sight, acts on geologic materials. Where it is the dominant agent, it produces a characteristic set of features. Because the primary action is solution, the effects are greatest where bedrock contains the greatest quantities of soluble minerals. Thus areas where limestone and gypsum are abundant are most likely to show the marks of ground water as a sculptor of landforms. Calcite is soluble even in pure water and is much more rapidly dissolved by water that is acid. Carbon dioxide is available from the atmosphere, and when in solution in water it makes carbonic acid; organic acids are present in soil as the product of decaying plants and animals. Downward-percolating water containing carbon dioxide and organic acids in solution moves into joints, along bedding planes, and through the more permeable beds of limestone, dissolving the rock as it goes. Once some openings have been enlarged, more water is diverted to them; in this way certain passageways eventually become much larger than others and form caverns of various sizes.

Figure 10–56 shows diagrammatically, in cross section, the conditions

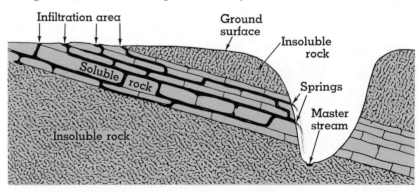

FIG. 10-56. Geologic setting favorable to solution of large caverns.

necessary for maximum development of solution caverns. Beds of lime-stone or other soluble rock dip toward a valley cut by a master stream. Infiltration of water occurs in the area indicated by arrows. Ground water then follows joints and beds both above and below the water table in three-dimensional networks of great complexity. Below the water table openings are enlarged in all directions, primarily by solution; though some enlargement may result from current action similar to stream erosion. Above the water table dissolved calcite is deposited in the caverns as water evaporates. As the master stream cuts deeper, more and more rock is drained and the water table drops, so that some part of the rock is always undergoing the change from saturation (and solution) to aeration (and deposition). Material deposited above the water table is known by the general term *dripstone*. The most common forms are *stalactites*, iciclelike bodies hanging from the roofs where there is leakage of water; and *stalagmites*, inverted cylindrical or conical deposits built up from the floor where water drips (Fig. 10–57).

Solution on a large scale produces changes in landscape that are most prominent in an area underlain by a thick, nearly horizontal sequence of limestone beds. Where two vertical joints intersect, a concentrated downward flow of water may dissolve more limestone than at other points. At the intersection, a depression may start at the surface, or a cavity may form at depth, and later collapse. The resulting depression is known as a *sink*. In some places sinks are found in straight lines, as if along a single large joint. They may grow and eventually merge into one long, narrow depression, termed a *solution valley*. Streams that go underground in a sink or solution valley are called *disappearing streams*; their water may reappear in the form of springs. A region underlain by soluble rock and characterized by some or all of these products of the action of ground water is said to have a *karst* topography, after the province of Karst, Yugoslavia, where the ground water processes were first studied.

The cycle of erosion in soluble rock is generally thought to take the following form:

1. In youth, most drainage is by surface streams, and subsurface drain-age is just becoming evident.
2. Maturity is characterized by the dominance of subsurface drainage over surface drainage and the lack of an integrated system of streams. The land surface shows many unconnected depressions and is irregular in appearance.
3. Old age is reached when most of the soluble rock has been removed and the drainage is again by streams.

The landscape in a mature karst region may resemble that of an area in which mass wasting is the dominant process. To identify an area of

FIG. 10-57. Stalactites and stalagmites.
(National Park Service.)

karst topography, one looks for soluble rock types, more-nearly-uniform lake levels resulting from better underground connections, and an irregular surface not dependent on the existence of a slope.

SUGGESTED REFERENCES ——————————————————

Davis, Kenneth S., and John A. Day: *Water: The Mirror of Science*, Doubleday & Company, Inc., Garden City, N.Y., 1961. (Paperback.)

Davis, William M.: *Geographical Essays*, Dover Publications, New York, 1954. (Paperback.)

Leet, D. J., and Sheldon Judson: *Physical Geology*, 3rd ed., Prentice-Hall, Inc., Englewood Cliffs, N.J., 1965.

Moore, G. W., and G. Nicholas: *Speleology: The Study of Caves*, D. C. Heath & Company, Boston, 1964. (Paperback.)

Rogers, John J. W., and John A. S. Adams: *Fundamentals of Geology*, Harper & Row, Publishers, Inc., New York, 1966.

Sharpe, C. F. S.: *Landslides and Related Phenomena*, Columbia University Press, New York, 1938.

Shimer, J. A.: *This Sculptured Earth, the Landscape of America*, Columbia University Press, New York, 1959.

Tolman, C. F.: *Ground Water*, McGraw-Hill Book Company, Inc., New York, 1937.

GLACIATION
AND WIND ACTION

CHAPTER **11**

DEFINITION AND GEOMETRY

A glacier is a mass of ice having definite lateral limits and motion in a definite direction, and originating from the compaction of snow. It may take a form that is basically linear, i.e., having one long dimension and two short dimensions; or sheetlike, having two long dimensions and one short dimension. Linear glaciers may be many miles in length and hundreds or even thousands of feet in width and thickness, and are generally found in valleys. Sheetlike glaciers may be several thousands of miles in length and width and a few thousand feet thick. The ice caps of Greenland and the Antarctic continent are presently existing examples. One type of glacier may grade into the other; for example, where glaciers from several mountain valleys flow onto a plain, spread out, and merge into an ice sheet.

ENERGY AND MOTION

The force of gravity is at work in both glacier formation and glacier motion. A load of overlying snow presses down on deeply buried snow and brings about its gradual compaction into ice. The ice, once formed, tends, as a crystalline solid, to flow plastically when sufficient force is applied to it. Downslope direction of flow dominates, and the horizontal component is strongest, but upward and downward flow also occur, as shown by the abrasive effects that may be seen on all surfaces of rock protuberances several feet high. Horizontal velocities range from so small as to be immeasurable to over 100 feet a day. Flow in a glacier is similar to that in a stream, the lowest velocities occurring near the contact of ice and the unmoving earth, and the greatest velocities at the center of the top surface.

MATERIALS

Ice is the principal material of glaciers, although meltwater may occur in intimate relationship with it, or as a significant accessory involved in transportation and deposition of clastic sediment in the area immediately adjacent to the ice. Rock fragments are incorporated in the ice and act as eroding tools.

As a crystalline solid, ice reacts to its environment according to the principles that apply to rock deformation (Chapter 6)—sometimes as a brittle material and sometimes plastically. When brittle, it fractures much as rock does. However, under sufficiently high pressure (at the bottom of a thick glacier; where it is pressed against an obstacle to its flow; or on the inside of a bend, where it is changing direction) ice may melt and move as water and then refreeze where the pressure is lower, or it may deform plastically. In these circumstances the glacier acts as a liquid with a very high viscosity.

UNIQUE PROPERTY

As a solid, ice has a number of characteristics distinguishing it from other geologic agents that carve out details of the landscape. The principal one is its strength. When water in joints freezes it may have enough tensile strength to pluck rock fragments from rest and cause them to move with the glacier. The fragments are held firmly, some pressing against adjacent rock surfaces and functioning as cutting tools to erode. A glacier flowing down a winding valley resists deformation and erodes more on the insides of bends than on the outsides, thus straightening the valley, unlike a stream. Because of its strength, particles of great size may be transported as readily as small ones; thus competence and sorting are of little significance.

LANDFORM PATTERNS

The erosional features of glaciation are in great measure parallel to the direction of flow, regardless of the shape and outline of the body of ice. But some are completely controlled by the shape and outline of the glacier, and their trends are parallel to or identical with the ice margins. The depositional features range from ice-deposited bodies of sediment parallel to the flow direction through ice-deposited bodies outlining the glacier, which may be transverse to the flow direction, to sediments deposited by meltwater on, in, under, or adjacent to the ice. Glacial topography may exhibit definite and characteristic patterns, may be highly irregular, or may resemble the results of other processes, such as mass movement, ground water, or wind. Because the surface is formed by ice, not by streams, and is irregular, surface drainage is not effective after melting of a glacier until sufficient time has elapsed for formation of a network of streams. The degree of integration of the surface drainage is at first low, then becomes higher the longer the time after glaciation has ended.

DISTRIBUTION AND EFFECTS

Glaciers now almost completely cover Greenland and the Antarctic continent and are found at higher latitudes and altitudes on all continents except Australia. About 10 per cent of the land surface of the earth is covered by glacial ice. In the geologic past great areas were covered by glaciers three times, the most recent ice age having ended only a few

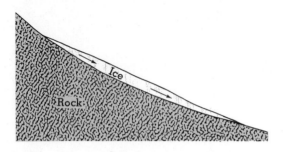

FIG. 11-1. Diagrammatic longitudinal section of a valley glacier at an early stage.

thousand years ago. Glacial action is of special interest to us because it is responsible for much of the world's most spectacular scenery—the great valleys of the Alps, South Island in New Zealand, and the fiords of the Scandinavian peninsula. Glaciers transported a thick layer of fragmental material equatorward, mixing a wide variety of materials and depositing them in warmer areas, where the weathering process formed soil of great agricultural productivity. Among the sedimentary products of glaciation are sand and gravel, so valuable for construction purposes and as a source of large quantities of ground water. The erosional effects of moving ice include the valleys and fiords previously mentioned and the scoured-out depressions now occupied by the Great Lakes of North America and many of the smaller lakes that abound in the adjacent areas.

GLACIAL FORMATION

The glacial process begins in an environment in which snow accumulates more rapidly than it wastes. The relative values of these two rates determine how rapidly and how large glaciers grow, but the existence of a glacier requires only an excess of accumulation. Temperature is important to the extent that the area must be above the *snow line*—that is, the altitude boundary above which there is permanent snow. The snow line is closer to sea level at high latitudes than at low, and at the higher altitudes at low latitudes. As snow accumulates in increasing thickness, it is compacted because of its own weight, the air between snowflakes is expelled, and ice crystals fuse to form granular material of greater density. When the total thickness is of the order of 100 to 200 feet, almost all the air has been forced out and an essentially solid body of ice exists at the bottom of the deposit, grading upward into snow at the top and modified there by melting and refreezing or by evaporation.

VALLEY GLACIATION

Valley glaciers show most of the characteristics of ice movement. Figure 11–1 shows diagrammatically and in longitudinal section the profile of a stream valley in which a glacier has accumulated. The glacier moves downslope under the force of gravity. It has a lower limit below which wastage (by melting and evaporation) exceeds the accumulation of snow plus the delivery by downslope motion. There is also an upper limit because as the glacier moves downslope it constantly removes ice, which must be replenished. As in many other geologic phenomena, the processes are balanced in a dynamic equilibrium, which is maintained by rapid adjustments to changes in the environment. In some areas changes may be visible from year to year, in others little variation is observed over long periods of time.

The glacier moves down the valley as a highly viscous body. The velocity may be measured by driving stakes into the ice and noting the rate of movement; speeds have been recorded ranging from a few inches to 150 feet per day. Velocity may vary from place to place in a glacier, for the rate of motion is greater where the valley gradient is greater and increases as temperature increases. Also, it may change temporarily because of disturbance by earthquakes or permanently because of an added load of snowfall accumulation or sediment piled on the glacier by mass movement originating on the adjacent valley walls. Solid ice flows down a sinuous valley and overrides obstacles in part as a result of deformation of crystalline matter under slow, long-continued application of force. But at earth-surface conditions ice has another property that is involved in flow: under pressure it liquefies without any temperature change. Figure 11–2 illustrates an experiment demonstrating this property. A beam of ice extends across the tops of two posts, and a thin wire with weights on both ends is looped over the beam. The force of the weights increases the pressure on the ice under the wire, and the ice melts; the water cannot support the wire and flows upward, refreezing above the wire, where there is no pressure. Similarly, as ice moves along a valley it is under pressure on the inside of bends and melts in reaction to the pressure. The water flows toward the outside of the bend, where the ice is in tension; it fills tensional cracks, and refreezes in the low-pressure zone. Thus glacial-ice deformation may occur either by this mechanism or by crystalline deformation.

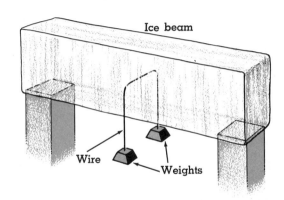

FIG. 11-2.   Pressure melting and refreezing ice.

Erosion.   Valley glaciers excavate material in a number of different ways. Rapid erosion occurs at the head of a glacier. Figure 11–3 shows in longitudinal section a *bergschrund* (from the German: *Berg*, mountain, and *Schrund*, crack), the space left at the head as the ice moves downslope and pulls away from the bedrock surface. Erosion occurs here partly through meltwater that cascades down into the bergschrund carrying solid particles that abrade and erode by impact, and partly through the hydraulic action of the moving water. Bergschrunds may be as much as several feet wide and tens of feet or more deep; hence large particles may be involved, with significant erosion resulting. At a time of freezing the bergschrund may become filled with ice, which connects the moving glacier with ice in joints in the rock. As the glacier moves down the valley it plucks out the blocks of rock that are surrounded by joints and are solidly frozen into the ice. By these processes the valley is eroded headward and the profile steepened, as indicated in the topographic profile of Fig. 11–3, which shows the preexisting stream profile by a dashed line. The head of the valley has been glacially eroded to a natural amphitheater called a *cirque* (Figs. 11–4 and 11–5). The steep, in some places even

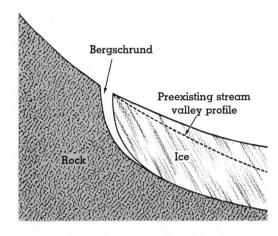

FIG. 11-3.   Head of glacier, shown in cross section.

vertical, slopes of cirques are favorites of mountain climbers for difficult climbs. Along a drainage divide, cirques may be so close to each other that the remnants of rock between them form peaks with extremely steep slopes, called *horns* (Figs. 11–4 and 11–6). The Matterhorn in the Alps and Mount Assiniboine in the Canadian Rockies are excellent examples of such landforms.

Rock material fallen to the glacier surface from the adjacent slopes or transported by streams or avalanches is concentrated along the edge of a glacier. In time and with snowfall it becomes incorporated in the ice. The fragments act as tools in the moving glacier, abrading the bedrock and leaving *striations* or deep *grooves* on the rock surface parallel to the flow direction. Much of the material thus removed is extremely fine grained, and is therefore known as *rock flour*. It causes the milky appearance of streams flowing away from glaciers. A high polish is often left on rock surfaces eroded by glacial action. Figure 11–7 indicates the typical distribution of material and the characteristic U-shaped cross section of glaciated valleys; Figs. 11–5 and 11–8 show such valleys in Alaska.

FIG. 11-4. Cerro Veronica, Peru, a glacial horn with two cirques on the near side. Bergschrund marked by crevasses in snow at heads of cirques and foot of cliffs on rock. (Aerial Explorations, Inc.)

FIG. 11-5.   U-shaped valley with cirque at head in distance. North Tongass National Forest, Alaska. (U.S. Forest Service.)

FIG. 11-6. Glacial horn, Glacier National Park. Note steeper upper slopes. (National Park Service.)

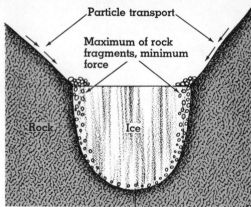

Particle transport

Maximum of rock fragments, minimum force

Rock   Ice

Minimum of large rock fragments, maximum force

FIG. 11-7. Valley glacier, shown in cross section.

FIG. 11-8. View looking upstream into a U-shaped valley eroded by a valley glacier. Cascade entering main valley from the left enters from a hanging valley. Ford's Terror Fiord, Alaska. (U.S. Department of Agriculture.)

EARTH SCIENCE

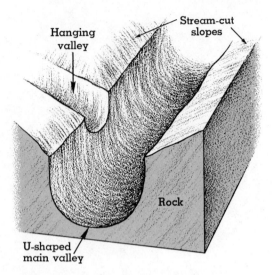

Hanging valley

Stream-cut slopes

Rock

U-shaped main valley

FIG. 11-9. Hanging valley.

The force with which the tools are pressed against the bedrock by the moving glacier is least at the surface and increases with the depth of overlying ice. One would therefore expect erosion to increase downward. Past a certain point, however, the tools have themselves been worn away and are therefore ineffective in deepening the valley. If we think of the rate of erosion as being related to the amount of abrasive material (the rock content of the glacier) multiplied by the force with which it is applied to the valley wall, the maximum product will be neither at the top of the ice nor at the bottom of the valley but at an intermediate point— which explains the U-shaped cross section cut in the originally V-shaped valley.

In the process of carving the valley, erosion is greatest on the insides of bends, and the valley is therefore straightened. Smaller tributary glaciers are less effective at downcutting than the main glacier because of their lesser thickness and the correspondingly reduced pressure. Ordinarily they join in such a way that the upper surfaces of the glaciers are concordant, or nearly so. Because of differences in the rate of erosion, valleys are deepened by different amounts, and the bases of glaciers may be at different elevations, as shown in Fig. 11–9. When the ice has melted, the stream flowing down the tributary valley drops to the level of the main valley in a waterfall. Such a feature is called a *hanging valley* (Fig. 11–8). Glacial valley bottoms do not necessarily slope downstream continuously, for where bedrock is locally more resistant the glacier can ride up over it. Depressions are likely to occur in such valleys.

Deposition. The material removed by a valley glacier through plucking, scratching, or gouging is transported and deposited by the ice. Some additional material handled by streams or mass movement may be mixed with the glacial sediment. All types of glacially deposited sediment are grouped under the general term *drift*. Material transported and deposited by ice, in contrast to meltwater, is unsorted and unstratified, and is called *till*. The fragments are likely to be angular (as opposed to rounded, stream-transported particles), with one or more flattened, striated, and abraded surfaces, and because of the comparatively low-temperature glacial environment, they are generally little weathered.

Moraines are deposits of till left by valley glaciers. As ice melts at the downslope end of a glacier, a great deal of the fine sediment is carried away by the meltwater. However, if the ice front melts back up the valley as fast as the ice body flows down, much of the contained clastic sediment is left where the ice melts at the glacier front. The moving ice may bulldoze a heap of sediment in front of it if flow rate exceeds melt rate, and a deposit of till is built up in the form of an irregular ridge across the valley.

Such a ridge, marking the furthest extent of the ice, is known as a *terminal moraine*. If during the warm and cold stages of climatic cycles the ice front alternately melts back and is immobile, morainal ridges are formed where the ice front has halted, forming a series of *recessional moraines* (Fig. 11–10). Debris also tends to accumulate along the edge of the glacier as it flows down the valley. After the ice melts, the material remains as a terracelike deposit of till called a *lateral moraine* (Figs. 11–11 and 11–12). A similar deposit that is formed at the junction of two glaciers and that continues down the central part of the valley is termed a *medial moraine* (Figs. 11–11 and 11–12).

FIG. 11-10. Emmons Glacier, Mt. Rainier. Ice near downslope end covered by rock debris. Trees are visible on the recessional moraine. Note deposition of valley train, or outwash, by the meltwater stream. (National Park Service.)

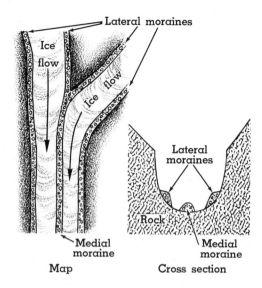

FIG. 11-11. Map and cross section of lateral and medial moraines.

Meltwater from the glacier flows down the valley, transporting and depositing fine sediment. These deposits are stratified and sorted; they are called *outwash* or *valley train* (Fig. 11–10). Except for the presence of an appreciable quantity of angular, flattened, and striated particles, and in their relationship to the moraines, outwash may be difficult to distinguish from the usual river-deposited sediment.

CONTINENTAL GLACIATION

Ice sheets, like valley glaciers, form where the rate of ice accumulation exceeds the rate of wastage, but the scale is a much larger one. Piles of snow and ice ultimately become so thick that, under the pressure of their own weight, they spread laterally, in equilibrium with the added load of snowfall, the ice temperature, surface slope, and other controls. In North America three major centers from which ice once flowed have been identified, one in Labrador, one in the Keewatin district adjacent to Hudson's Bay, and the third in the North American cordillera. From the centers, ice flowed radially in lobes that spread overland, now one, now another moving farther, often with adjacent lobes differing greatly in their direction of flow. Recession followed by advance left striations superimposed in set upon set, showing flow directions as much as 90° apart in successive advances of the ice.

Erosion. Erosional features include many that also characterize valley glaciation—striations, grooves, and polished rock surfaces, for example.

Deposition. Because of the difference in size and shape, there are some important contrasts between the depositional features of valley glaciers and continental glaciers, but the meaning of the terms drift, till, outwash, terminal moraine, and recessional moraine is the same. Because of the gigantic size of the continental glaciers and glaciated areas, individual deposits are very large. Outwash deposits may be scores of square miles in area and over 100 feet thick. Terminal and recessional moraines may make belts of hills many miles long, several miles wide, and with relief of several hundred feet. *Ground moraine* is the term applied to the comparatively thin layer of till left in the glaciated area after the ice has melted. Whereas terminal and recessional moraines are extremely hummocky and irregular in surface conformation, slopes in the area of ground moraine are usually more gentle, the glacially smoothed surface dominating over the irregularities created by deposition.

Drumlins, elongated hills that usually have one end broader and more steeply sloping than the other, occur in parallel arrangement in swarms in some areas of ground moraine. Excavations show barely perceptible stratification parallel to the surface, and predominantly clay content with

admixture of small amounts of coarser sediment. The broader and steeper end faces the direction from which the ice came. Drumlins reach dimensions of 2 miles in length, ¾ mile in width, and several hundred feet in height. It is thought that they were formed by the plastering of clay, layer by layer, on raised areas beneath the moving ice, for their orientation is clearly related to the motion of the glacier, and some of them have bedrock cores. The drumlins provide one basis for estimating the thickness of the overriding ice sheet. The naturally occurring material is densely compacted and difficult to excavate. In the laboratory the clay can be broken down by mixing with water, and can then be compacted under pressure until it reaches the same degree of compaction as the sediment in place. The pressure required to achieve this is found to be equivalent to a thickness of ice of the order of 10,000 feet. Drumlins, recessional moraines, and terminal moraines are so clearly oriented that they give a pronounced and unmistakable fabric, or directional pattern, to the surface in many areas of continental glaciation.

Erratics are large, glacially transported boulders scattered throughout the drift. They are usually of a rock type different from the local bedrock, thus testifying to the fact that they were transported from another area. Often the specific source region can be identified. In Europe and North America many *boulder trains* of erratics have been found in lines or fanshaped patterns extending a number of miles from the outcrop, clearly demonstrating the flow direction of the ice sheet.

Stratified and sorted sediment was deposited by meltwater from the ice sheet. Extensive bodies of outwash (Fig. 11–13) are found bordering terminal and recessional moraines on the side not covered by ice. Meltwater flowing off the glacier and emptying into water that was ponded between the ice front and a terminal or recessional moraine deposited sand and gravel in comparatively small units. Meltwater streams flowing in channels on top of the ice or through tunnels in or beneath the ice left deposits of bedded sand and gravel, called *eskers*, which now form sinuous, elongate, narrow ridges, some as much as 23 miles long, 120 feet high, and 500 feet wide. Similar sediments, *crevasse fillings*, were deposited in *crevasses* or fissures in the ice by meltwater.

Many other frequently seen depositional features are of glacial origin. A *kettle* is a depression caused by deposition of sediment, usually outwash, around a stranded block of ice that subsequently melted. It may be irregular and steep sided. Lakes are abundant in glaciated areas, and the fine sediment in them has characteristic laminations of alternating light-colored silty clay and dark-colored organic clay, called *varves*. The alternating bands are thought to represent summer and winter deposition, respectively. In summer seasons flow rates and competence were high, temperatures high, and oxidation more complete; thus the thicker, light-

FIG. 11-13. Stratified and sorted glacial outwash, central Michigan. (U.S. Department of Agriculture.)

FIG. 11-14. Moraines, Amenia, New York. Irregular low hills in foreground are typical of morainal landforms. (The American Museum of Natural History, New York.)

colored layers of coarser sediment were deposited then, and conditions were reversed in the winter. Sequences of varves are like tree rings in that it is possible to recognize similar patterns of thickness for hundreds of laminae in sediments of lakes close to each other. As the ice front retreated, new lakes were formed, and varves deposited in them in overlapping sequences have been correlated over large areas. Since a pair of laminae represents one year's deposition, it is possible to estimate approximately how many years were involved in the retreat of the glacier front in a specific area. European studies indicate the period since the last glaciers retreated there to be about 16,000 years.

Because the area of ground moraine is essentially one of erosion, it is generally lower in altitude than the recessional moraines, terminal moraines, and outwash, which are depositional features (Fig. 11–14). Outwash plains are lower than recessional or terminal moraines, and, except for kettles, usually have very smooth surfaces. The recessional and terminal moraines are extremely irregular in relief and lack topographic pattern except that they form broad belts extending cross country and outlining the ice lobes. In clayey sediments of moraines all the drainage is by streams. Because of the irregularity and the numerous closed depressions, lakes and swamps are present in great numbers in areas recently glaciated. Many streams empty into lakes without surface outlets, and the drainage pattern is nonintegrated. But as time passes, connections between the separate systems become established and the surface drainage is integrated into a single system. Thus the degree of integration of streams in a glaciated area is an indication of the length of time that has passed since the retreat of the glacier.

WORLDWIDE CONDITIONS ACCOMPANYING AND FOLLOWING GLACIATION

The increase in snow accumulation that caused the glaciers of the last million years is only one indication of a global change to colder climates. Evidence indicates that elsewhere in the world there were lower temperatures and corresponding changes in vegetation during the ice advances, and also, but to a lesser degree, until retreat of the ice was nearly complete. Precipitation generally increased, so that extensive lakes and rivers once existed in the now arid regions of parts of Nevada and Utah and of the Sahara in North Africa. Shoreline features of large lakes and abandoned valleys of rivers the size of the Rhine are found in northern and western Africa, also bones of animals of semiaquatic habit. The Great Salt Lake of Utah is a remnant of Lake Bonneville, a vast lake that once occupied many of the valleys of western Utah and Nevada.

The tremendous weight of ice on the continents for long periods of time caused depression of the crust in those areas. Since the melting of the ice, the crust has rebounded isostatically. Evidence of this is found in the tilting of shoreline features of postglacial lakes in glaciated areas. The amounts of uplift have been calculated and the values contoured. The pattern of contours supports the ice-center location established by other evidence. Upward movement is still continuing, because the strength of the crust and the slow shift of material in the mantle do not permit instant adjustment. Rates of uplift have been measured in many places in Europe and North America. The pattern of uplift is such that in time the Detroit River, between Lake Huron and Lake Erie, may be the highest point in the St. Lawrence waterway, and Lakes Superior, Huron, and Michigan will drain through the Illinois River into the Mississippi River rather than through the St. Lawrence River.

The withdrawal of water from the oceans, and its storage on the continents as ice, lowered ocean-water levels around the world. Numerous estimates have been made of the changes in sea level. A reasonable value is 300 feet—sufficient to establish land bridges connecting the British Isles to Europe, and Asia to North America, and to expose great areas of what is now shallow sea bottom around the present continents. Predictions of sea-level changes if present glaciers were to melt range from 100 to 200 feet of rise. The number of people living below this altitude is very great, and the value of property and agricultural and industrial production involved is incalculable. The future temperature regime of the earth thus becomes a matter of great interest. To predict the course of events we must understand what has happened to date, a matter to which we now turn our attention.

CAUSES OF GLACIATION

Many theories have been proposed to account for continental glaciation, some reasonable and some less so. Any theory must take into account the following established facts:

1. Worldwide glaciation has been a rarity.
2. Most of the recent glaciation was in the Northern Hemisphere.
3. There were four periods of glaciation that occurred concurrently throughout the world.
4. The ice melted off rapidly at the end of the glacials (time of glaciation).
5. The climate of at least one interglacial (time between glacials) was warmer than the present climate.

It would be of value to know whether we are now in an interglacial, with a glacial interval approaching, or if the glaciation has ended for a period of some millions of years.

Of the many theories of glaciation that have been proposed, a few have the widest acceptance, but there are some objections to all of them. Nevertheless, it is of interest to review those that have been most generally considered.

Crustal uplift. If the land surface were to be raised above the snow line, glaciers would form and spread into the adjacent territory. We know that the crust is mobile, that areas of continental size can move vertically, and that upward motion can amount to tens of thousands of feet; consequently such uplift appears to be possible. However, there are major objections to such a hypothesis: (1) The last glaciation, with its four advances separated by interglacials, would have involved larger sections of the crust moving upward and downward more rapidly than at any other time in earth history. (2) The earth is so old that in general all processes of the scope of continental glaciation have been repeated with but minor variations not once but several times; the single exception is multiple glaciation like that of the recent glacial epoch. (3) The climate during the interglacials was warmer than it would have been if it were caused by uplift and subsidence alone, so we must conclude that some other control must have been important.

Displacement of the poles. We observe that ice caps now exist only near the poles of the earth. Is it not possible that the axis of rotation of the earth, and therefore the positions of the poles, has shifted with respect to the continents so that the North Pole was once near Hudson's Bay? Subsidence of the crust in one place and uplift in another could change the position of the center of gravity and the orientation of the earth's axis of rotation. Subcrustal convection currents could cause large displacements of the surface with respect to the interior. One-time continental polar areas are now at lower latitudes, by this reasoning.

A major objection to this theory is the statistical impossibility of four axial, polar, or crustal shifts occurring as rapidly as would be necessary and to almost the same position each time. Another difficulty is to explain why the present centers of glaciation are so near the centers of the glacial maxima. This suggests strongly that if polar wandering occurred at all, some other control was stronger.

Continental drift. This hypothesis, discussed earlier (Chapter 8) in connection with theories of mountain building, has substantial evidence in its favor. Continental drift has been offered as the explanation for the widely separated concurrent glaciations of the different continents, on the assumption that the original continent was polar in position before it broke up. But this in itself does not explain the cyclic glaciation, and the evidence indicates that the drift apart occurred long before glaciation began.

Atmospheric control. The amount of solar radiation received at the earth's surface controls temperature conditions and other weather elements and consequently the position of the snow line. One cause of glaciation that has been suggested is a variation in the atmospheric content of carbon dioxide. A high content causes the "greenhouse effect" (Chapter 2), which traps heat in the lower atmosphere and raises temperatures there. Presumably a lower carbon dioxide content would have the reverse effect and could thus set off an interval of glaciation. The fluctuation of the carbon dioxide content of the atmosphere has been proposed to explain other phenomena of the geologic past, but there is no corresponding record of glaciation, so carbon dioxide fluctuations may not be significant. There is also a large body of evidence indicating a long-time constant atmospheric composition.

A second atmospheric control of solar radiation received at the surface of the earth is dust content. Solid particles in the air reflect radiation out into space and when present in sufficient quantity can affect weather and climate. One source of large amounts of dust is explosive volcanic eruptions. Krakatoa, a volcano in the East Indies, in 1883 put enough volcanic dust into the air to affect the weather over much of the world to the extent of reducing world temperatures and presumably increasing precipitation. It has been suggested that vulcanism on a larger scale and for longer periods than now observed was responsible for the glaciation of the past. However, two objections difficult to override have been raised. First, at other times in the geologic past vulcanism in North America provided thousands of cubic miles of pyroclastics distributed over hundreds of thousands of square miles of area and continuing throughout several millions of years without being accompanied by glaciation. Second, we do not find pyroclastic sediment indicating large-scale explosive vulcanism at the times glaciation was widespread.

Solar cycles. Variable stars are well known to astronomers, and it has been suggested that varying radiation from our sun has caused the cold glacials and warm interglacials. However, there is much evidence to indicate that solar radiation has been uniform, at least since life originated on the earth. Man has been observing the stars for a relatively short time and the data available are probably not sufficient to provide a sound basis for any hypothesis involving the rate of solar-energy emission. The cycle involving radiation, evaporation, atmospheric circulation, and precipitation represents a sensitive equilibrium, and it is possible that small changes in solar radiation may have greater effects than we can predict with assurance in our present incomplete state of knowledge.

Astronomical configurations. Glaciation, variations of ocean level, and oceanic climate fluctuations in the last half-million years appear to be closely related to the rate at which energy has been received from the sun and its distribution over the earth. The significant controls are the periodic tilt of the earth's axis, precession (rotation) of the axis, and changes in the eccentricity of the earth's orbit. These variations, when in the proper combination, cause a minimum of energy to be received in the higher latitudes where snow accumulates and maximum energy in the lower latitudes where evaporation occurs, and could thus cause glaciation. Warm interglacials would occur when the relationships were reversed.

Sea-level ocean-circulation control. A theory of glacial origin was proposed by the American geophysicists Maurice Ewing and William L. Donn that takes into consideration multiple characteristics of the earth rather than any single relationship. They suggested that sea-level control of oceanic circulation has affected climatic patterns to the extent of causing continental glaciation. The Arctic Ocean is connected to the Atlantic and Pacific Oceans by comparatively shallow, narrow bodies of water. A sea level higher than the present one would enlarge the connections, and warm currents would enter the Arctic Ocean, evaporation there would increase, atmospheric moisture would be greater, and precipitation would increase. At that latitude the moisture would fall as snow, and continental ice sheets would form, spreading equatorward until enough water was removed from the oceans and stored on the continents as ice sheets to lower sea level to the point where warm-water circulation into the Arctic Ocean would again be shut off. Water temperatures would then drop, evaporation and precipitation would be diminished, the ice sheets would waste away, water would be released in the process, sea level would be raised, and the stage would be set for repetition of the cycle.

The simplicity of this hypothesis is attractive. The cyclic pattern would explain the rapid oscillation between the extremes of glacials and interglacials. A particular set of interrelationships in the polar sea and bordering oceans and continents is required, but it is reasonable to assume that such an environment favorable for continental glaciation might have occurred only occasionally, thus explaining the relative rarity of the phenomenon.

THE WIND

In the arid and semiarid regions of the world, the wind is an active and effective agent of erosion and deposition. This does not imply the absence of other processes, for streams and mass wasting are also significant, if not dominant—but the unique effects of the wind are added to or superimposed on landforms produced by other agents. It is difficult to evaluate precisely the comparative importance of stream and wind action in arid regions. The processes are generally slow acting, and the angular desert landforms give the appearance of an inactive environment. Precipitation is characteristically very irregularly distributed in both time and space, and the changes in the topographic surface of stream origin, though distinct, occur less often than those caused by wind, which acts more or less continuously in the long intervals between rains. The effectiveness of wind action in dry areas is due to the absence of a stabilizer such as vegetation, which protects soil from erosion by frictional reduction of wind velocity near the surface, binds the soil together with a mat of roots, and covers the soil with organic litter. Similarly, sand and silt when moist can be eroded by the wind only with great difficulty, but when they are dry, no restraints exist and surprisingly large amounts of clastic sediment may be eroded rapidly. If water is abundant and near the surface, it acts both by binding the soil together and supporting vegetative cover. Under unusual conditions evidence of wind action may be found even in humid regions.

The wind carries clastic sediment in exactly the same way that water does, and the settling velocity of particles is just as important in the air as in water. The finest particles may be carried in suspension thousands of miles from the point of origin before settling out of the air. But the atmosphere is unlike streams in that there is no distinct upper surface, and winds are seldom linear in form. Thus there is no upper limit to the heights at which dust can be carried other than the tropopause, the upper limit of vertical air currents, which may be as much as 12 miles above sea level. Actually, most suspended dust is transported in the lower few thousand feet of air, and coarser particles move in saltation near the surface.

The wind has a much smaller competence than streams; because of this, wind-blown sediment is extremely well sorted. Sediment in saltation is near the surface, and the zone of vertical transition from much sand to no sand is quite thin. In one sandstorm in the Egyptian desert, the zone of sand in saltation extended from the ground surface to a line about midway up the windows of the car used by a geological field party. The sand removed the paint on the external metal surface and frosted the window glass up to the 1-inch transition zone in which the sandblast effect graded from complete to nothing. Sediment also moves in traction, but again—as with sediment in saltation—if any sediment is in motion, a great deal is in motion, and such a flow of material is difficult to study.

GEOMETRY OF WIND ACTION

Near-surface winds occur in broad areas within which direction or velocity of movement or both may remain relatively constant for various periods of time. Winds may converge on a moving center or diverge from one with a spiral flow pattern. Because the atmosphere is many miles thick and air motion involves a large vertical component, winds may gain or lose many thousands of feet of altitude within a few miles of horizontal motion.

ENERGY AND MOTION

Gravity acts to provide energy and produce motion of the air to some extent, but most winds are established directly by a phenomenon that does not affect other geologic agents: differences in air pressure (see Chapter 3). Air is free to move in the vertical direction more than any other geologic agent. By far the greatest amount of air moves in paths roughly parallel to the earth's surface, ranging in relief to the same extent as the topographic surface. Velocities range from zero to very great.

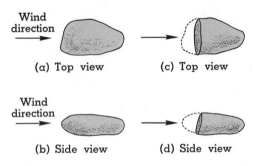

Wind direction →

(a) Top view

→

(c) Top view

Wind direction →

(b) Side view

→

(d) Side view

FIG. 11-15. Pebble (a and b) eroded to form a ventifact (c and d). Natural size.

MATERIALS AND SIGNIFICANT PROPERTIES

The air is a gas and is therefore not restricted in vertical distribution. It has a low specific gravity, and thus has less flotation effect on clastics than water, the other significant geologic fluid. Its viscosity is low compared with that of water, so particle settling velocities are comparatively high. Sorting of particles takes place, and sediment deposited by the wind is layered. Solution does not occur. Because the wind is ordinarily of low competence, particles in motion are normally of sand size or smaller. The effects of abrasion and polish are visible on larger solid bodies, in some places selectively.

FUNDAMENTAL PRINCIPLE

The settling velocities of solid particles in air control erosion and deposition of sediment by the wind. Wind action is similar to stream action except for the absence of erosion by solution and the lack of restriction in the vertical and lateral directions. As winds blow across even the highest mountain ranges, fine particles held in suspension may be carried from one intermontane valley to the next, as if there were no mountain barrier between them. Similarly, there is less topographic control of deposition by the wind, and air-transported sediment may be deposited in a blanket of nearly uniform thickness on mountain tops and in valleys.

LANDFORM PATTERNS

If wind directions are constant, both erosional and depositional landforms of wind origin take on a high degree of parallelism. With changing directions, the pattern becomes less apparent, to the extent of a complete lack of orientation. Because of the wind's freedom to act in the vertical direction, depressions can be maintained on all scales without connection, and surface drainage patterns are likely to be substantially nonintegrated.

WIND–FORMED FEATURES

Features produced by wind as a geologic agent range in size from the well-rounded sand grain with a surface frosted by many impacts to depositional landforms many miles in maximum dimension. Like those of other origins, they may be grouped into erosional and depositional categories.

Erosion. Small products of wind erosion are *ventifacts*, gravel particles that have been abraded by the sandblast of a wind of uniform direction. A unidirectional wind removes material from the windward side of a particle, as shown in Fig. 11–15. Arrows show the wind direction and views (c) and (d) represent the pebble after the material within dashed lines has been removed by abrasion. The smooth, gently curved, abraded surface faces the wind source. Multiple faces may be formed by consistent wind directions in sequence or by shifts in the position of pebbles at long intervals. A *blowout* is an area of excavation of soil by the wind, usually where a stabilizing agent is lacking. Blowouts range in size from pits a few inches in diameter and depth to the Big Hollow, a depression in the Laramie Basin, Wyoming, which is 9 miles long, 3 miles wide, and 150 feet deep.

If gravel is present in the soil, the coarse particles accumulate at the soil surface as the sand and finer particles are removed by wind erosion, and they may form a layer one particle thick with pebbles as closely and neatly arranged as in a mosaic. Such a layer is termed a *desert pavement*. Often the pebbles are impregnated with iron oxide released in the weathering process, and the exteriors are polished by the moving fine sediment to a smooth surface called *desert varnish*. Cemented bedrock may be unequally eroded to an irregular surface of some relief.

Deposition. Clastic sediment transported and deposited by the wind may form either hills, called *dunes*, or a more-or-less uniformly thick layer over a preexisting surface.

Dunes are composed of sand-size particles, usually quartz, less often feldspar, and rarely calcite or gypsum. Elements required for their construction include a supply of sand, a constant or nearly constant wind direction, and the lack of a stabilizing agent. Sand sources may be the soil in arid and semiarid regions; certain bedrock types that weather mechanically to produce large amounts of sand; alluvial sediments in floodplains; shoreline sands of lakes and oceans; and glacial outwash. Sand dunes grow, move, and have characteristic forms, depending on wind direction. If the wind is unidirectional, the dunes will be nearly uniform in size and shape. A changing wind direction causes modification or destruction of form (Fig. 11–16) and affects the geographic distribution of the dunes. Dunes can neither grow nor move if they are stabilized by a cover of vegetation, and, in fact, planting may be undertaken in order to halt the progress of dunes. In some areas of the world where vegetation has increased, usually as a result of climatic change, many large dunes occupying extensive areas have been stabilized. The material and form of the dunes are still recognizable, but their growth and migration have ceased.

Like so many other geologic processes, the growth of dunes is not completely understood. They are believed to form by a first accumulation of sand in the lee of some surface irregularity. There the air velocity and competence are low, and sand settles to the ground surface as snow drifts on the leeward side of a fence, a bush, a building, or other obstruction. Once the sand accumulation is initiated, it continues by accretion on the

FIG. 11-16. Ripple marks on sand dunes, varying wind directions.
Death Valley, California. (National Park Service.)

EARTH SCIENCE

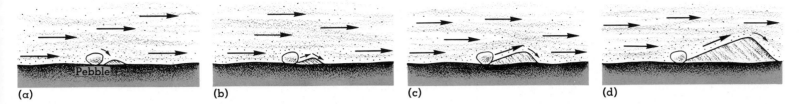

(a) (b) (c) (d)

Pebble

FIG. 11-17. Growth of a sand dune by deposition in the wind shadow of an obstruction. Arrows show velocity and direction of wind.

leeward side. Growth is thought to occur as shown in Fig. 11–17. The wind is blowing from left to right, and sand accumulates on the leeward side of the pebble. As time goes on, sand is blown up the gentle windward slope (the steepness related to sand size and wind velocity), falls over the crest of the dune, and slides down the leeward surface, which is at a nearly uniform slope of 32 to 34° from the horizontal (Fig. 11–18). This

FIG. 11-18. Leeward slope of sand dune, showing angle of repose. Great Sand Dunes National Monument. (National Park Service.)

GLACIATION AND WIND ACTION

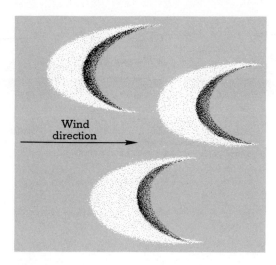

FIG. 11-19. Map view of barchan dunes.

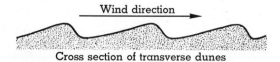

Cross section of transverse dunes

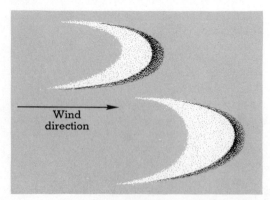

Map view of parabolic dunes

FIG. 11-20. Transverse and parabolic dunes.

is the angle of repose, the slope assumed by equidimensional fragmental material piled in the air. The sand will be laminated, and the laminae are preserved leeward slopes. Sand on the lee slope is loose, that on the windward slope comparatively firmly packed.

Dunes appear in many forms, as viewed from above. If the supply of sand is small and the wind direction constant, *barchan* dunes form; they tend to be crescentic and are separated one from another, as shown in Fig. 11–19. If the supply of sand is large and a layer of sand many feet thick is present everywhere, the dunes will take a different form. *Transverse* dunes (Fig. 11–20) are oriented with their crests about perpendicular to the wind direction. They occur in great numbers, with similar orientations resembling waves upon the ocean, except for the asymmetry of the dunes. The *parabolic* dune (Fig. 11–20) is representative of many dune areas. Since the mechanics of dune formation are not completely understood, some of the reasons for their diversity in size and shape have yet to be discovered.

Fine sediment (clay and fine silt) can be transported thousands of miles in suspension. When it settles out of the air it forms a blanket of more-or-less uniform thickness covering the preexisting surface regardless of slope. This deposit is known as *loess*. It is composed of angular, well-sorted, loosely packed, and uncemented particles of various minerals. It has a vertically prismatic structure similar to the structure of a mature soil. Characteristic features of loess are its tendency to maintain vertical cut slopes for many years but to erode rapidly on more gentle slopes, and a strong tendency to compact if a load is placed on it and it is wetted. Many structures built on loess have been severely damaged by uneven settlement after long rainy spells. In North America, extensive and thick loess deposits exist in an area in the central United States that is peripheral to the area covered by continental glaciers. It is thought by some geologists that winds blowing from the center of the ice sheet toward the adjacent warmer land area transported fine clastics and deposited them there. Others believe that the loess deposits originated in the weathering bedrock and soil of the high plains just east (and leeward) of the Rocky Mountains. Comparable deposits in China are believed to have originated in the interior desert of Asia.

The wind is free to move up and over topographic highs and then drop down into the basins in between, eroding and depositing as it moves. In this respect it is unlike water, which acts either downward or laterally and most of the time follows such paths that each tributary enters the larger stream at the elevation of the stream. Areas in which the wind is a significant gradational agent are likely to have isolated topographic basins not integrated into the regional drainage system; and the landforms, both positive and negative, are likely to occur in irregular patterns,

300

as seen from above. Even if there is a vegetative cover and action by other agents of erosion and deposition, the irregular patterns linked with well-sorted, sandy materials are usually easily recognized as being of wind origin.

Recent activity by oil companies in the arid regions of the Middle East has necessitated intensive study of the motion of wind-blown sand. Housing, petroleum pipelines, refineries, and other structures have required protection from burial or erosion. The problem has been at least partly solved by construction of gigantic "sand fences" on the windward side of the affected area, deflectors to bypass the sand around the structure to be protected, or paved areas to speed it past them.

SUGGESTED REFERENCES

Flint, R. F.: *Glacial and Pleistocene Geology*, John Wiley & Sons, Inc., New York, 1957.

Leet, D. J., and Sheldon Judson: *Physical Geology*, 3rd ed., Prentice-Hall, Inc., Englewood Cliffs, N.J., 1965.

Muir, John: *Yosemite*, Doubleday & Company, Inc., Garden City, N.Y., 1962. (Paperback.)

Pound, Elton R.: *The Physics of Ice*, Pergamon Press, Inc., New York, 1965. (Paperback.)

Putnam, William C.: *Geology*, Oxford University Press, New York, 1964.

Schultz, Gwen: *Glaciers and the Ice Age*, Holt, Rinehart and Winston, Inc., New York, 1963. (Paperback.)

Shapley, Harlow (ed.): *Climatic Change: Evidence, Causes, and Effects*, Harvard University Press, Cambridge, Mass., 1954.

Shelton, John S.: *Geology Illustrated*, W. H. Freeman & Co., San Francisco, 1966.

COASTAL LANDFORMING PROCESSES

PROCESSES

CHAPTER 12

12

Most aggradational and degradational processes on the coasts of lakes and oceans take place in a linear zone of small vertical extent. The resulting landforms are usually found very near the shoreline, either below or above water. However, because of the mobility of the earth's crust, the fluctuation of sea level in the course of geologic time, and the comparatively short life of even the largest lakes, we find evidence of ancient shorelines at sites far from existing bodies of water. For example, much of the evidence for long-extinct glacial and postglacial lakes consists of "fossil" erosional and depositional landforms associated with beaches and shorelines.

In much of the discussion to follow we shall refer to the oceans, but what is said applies equally well to lakes, although on a smaller scale.

WAVE ACTION

GEOMETRY

The zone of greatest erosion and deposition by waves is essentially linear, following the shorelines of lakes and oceans, and is regular or irregular depending on the shape of the land surface. Tides, storms, worldwide changes in sea level, and deformation of the crust cause the shoreline to move both horizontally and vertically, but such changes in position are small in comparison to the total length of the shore.

ENERGY AND MOTION

The source of energy for the normal deep-water wave is the wind, blowing across the water surface and disturbing it so that waves are formed. It is the wave form that moves, not the water. Waves proceed from the source to adjacent areas, continuing until their energy is dissipated, even though the wind may have stopped or left them behind. Thus there may be waves at a point where no wind is blowing. Wave velocities may reach several tens of miles per hour.

When a wave moves into shallow water the crest moves more rapidly than the water at the bottom, where friction acts to retard motion. Lateral movement of the water itself then occurs. Waves overtop and collapse, or "break," and the turbulent water in the breaker zone disturbs bottom sediment. The water rushing shoreward may carry sediment with it and then return seaward with sediment in transport. Erosion at the shoreline may be by abrasion, solution, or the impact of water on solid materials.

MATERIALS

Water in waves acts like water in streams, eroding clastics to the limit of its competence and dissolving solids to the extent of their solubility. The principles governing these processes differ little from those that apply to streams.

FUNDAMENTAL CONCEPTS

Wave action is restricted to a narrow zone adjacent to the water's edge, departing little from this restricted area either horizontally or vertically. Erosional and depositional forms that are the result of wave action occupy a similarly restricted part of the land area. The repeated impact of waves on the shore, the direction of wave attack (whether perpendicular or oblique to the shoreline), and the two-directional motion of water and sediment are the most important features of this environment. The resulting landforms may be lifted above the reach of further wave action or depressed below it by crustal or water-level changes and may then be modified or destroyed by agents existing in the new environment.

LANDFORMS

Landforms along a shoreline are both erosional and depositional. Both types are linear in map view and are parallel or subparallel to the water's edge. In each type the new land surface is nearer the water level than the old, resulting in formation of a *bench*, or terrace, with its surface near sea or lake level, which has been cut down by erosion or built up by deposition. When, as often occurs, the water level drops relative to the land, one or more elevated terraces remain some distance inland, above the water level and roughly parallel to the new water-land boundary.

WAVES

In all geologic processes, energy must be available if work is to be performed. Erosion and deposition occur in the shore environment through the energy supplied by wave motion and currents established by waves. Wind blowing across the surface of a body of water forms waves when its velocity is greater than 2½ miles per hour. Once initiated, the waves become larger until they reach a maximum size that is related to the particular wind velocity. The greater the distance traveled under the effect of a wind (the *fetch*) the larger the waves become—which suggests the reason for the greater size

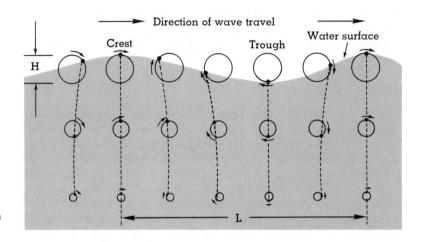

FIG. 12-1. Relation between wave motion and water motion, shown in cross section.

of the swells that reach the California coast as compared with those on the Atlantic shore.

In deep water the wave form moves laterally, but the motion of an individual molecule of water is quite different. The succession of waves forces the water molecules to follow a roughly circular path, as viewed from the side, and they do not move appreciably beyond this limited orbit during the passage of a wave. Figure 12–1 shows the motion of representative water molecules at different depths as a wave passes from left to right. The diameter of the circular path decreases with increasing depth. Displacement is small at great depths, but evidence of wave motion has been found in sediments as deep as 800 feet. Wave profiles are symmetrical, as shown in the figure. The terms used to describe waves are length (L), height (H), and velocity (V). Length and height are the distances horizontally from crest to crest and vertically from trough to crest, respectively. Water depth can be reduced to a minimum of about ½L without appreciably affecting wave characteristics, but in shallower water wave lengths decrease, heights increase, and velocities are reduced.

WAVE EROSION AND DEPOSITION

The energy applied to materials on shore by wave action may be very great. Mechanical effects are generally accomplished by impact and abrasion. During one storm in the North

Sea, waves up to 42 feet in height and 500 feet in length struck the shore of a port in Scotland with impacts as great as 3 tons per foot per wave. A concrete seawall was broken up, and one fragment weighing about 2600 tons was moved seaward.

The motion of particles causes them to abrade the shore and bottom materials and to become rounded themselves. Angular fragments of granite have been known to become rounded within one year. The impact and momentum of water may be converted into hydraulic pressure if the water strikes jointed rock. The pressure transmitted through water and air in the joints may be great enough to dislodge blocks of rock, which are then subject to abrasion, and to solution if the rock is soluble. Water is generally a better solvent of limestone at the surface than at depth, and a landform known as a *solution notch* (Fig. 12–2) is found in many places where limestone is exposed on the shore.

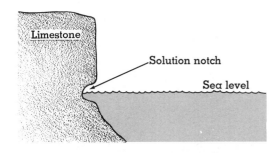

FIG. 12-2. Cross section of solution notch formed in limestone near sea.

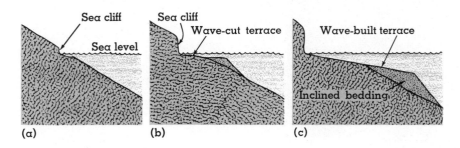

(a) (b) (c)

FIG. 12-3. Cross section showing erosional and depositional features that result from wave action (vertical highly exaggerated).

FEATURES OF STEEPLY AND MODERATELY SLOPING BOTTOMS

Figure 12–3(a) shows a steeply sloping land surface and bottom at a shoreline. Waves strike the rock or other material, removing it in the zone of most active wave action, undercutting the bank, reducing particle diameter, and removing the finest particles as the water returns seaward. A steep *sea cliff* (Figs. 12–4 and 12–5) is formed, with the slope determined by the properties of the material: overhanging if in rock with joints dipping landward, sloping seaward at about 30° if in uncemented sand. A gently sloping bottom, the *wave-cut terrace*, shown in Fig. 12–3(b) and (c), is also formed, with depth and slope depending on the wave dimensions, properties of the material, particle size, and other factors. The controls of the surface form are complex, for it is both an erosion surface and a transportation surface. After it has been cut, it affects the characteristics of the waves that cross it, further complicating the interrelationships. Thus the development of a wave-cut terrace is not yet completely understood.

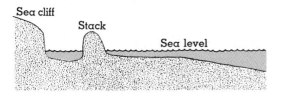

FIG. 12-4. Cross section of a stack.

FIG. 12-5. Sea cliffs eroded into extrusive igneous rocks, Hawaii. (National Park Service.)

The material eroded by wave action may be transported seaward once the particle size is small enough, and will be deposited at a depth below the level of wave erosion. The wave-cut terrace (on the bedrock or other material) is continued as a surface on the deposited sediment [Fig. 12–3(b) and (c)]; this new surface is called a *wave-built terrace*. Sediment is transported across this surface also. The sediment is laminated, as shown in the figure. If the bedrock is unequally resistant to erosion, a body of rock may become isolated from the mainland. Figure 12–4 shows such a remnant, known as a *stack*.

Wave energy is rapidly reduced as waves move across a shallow, gently sloping bottom. For this reason, it is believed that, without a relative rise in sea level or subsidence of the land, there is a maximum possible combined width of the wave-built and wave-cut terraces, depending in part on the properties of the material involved. Extensive coastal terraces are thought to occur only where there is a rise in sea level adjusted to the rates of erosion and transportation.

On an irregular coastline, headlands are actively eroded while the intervening embayments are not only little eroded but are the sites of sediment deposition. The difference results from wave refraction related to the depth of water. Figure 12–6 shows the features of this process. Solid lines are advancing wave crests and dashed lines indicate the paths of energy. Note that the dashed lines are uniformly spaced on waves *a*, *b*, and *c*, which are in deep water, indicating that energy distribution along these waves is uniform. As the waves approach the shore, they are retarded where the water shoals near the headland, the wave crests in deeper water get ahead, and the crests swing toward the headland. Energy travels in the direction the waves move (perpendicular to the crests); thus the closer spacing of the dashed lines at the headland indicates that erosion is more vigorous there. This nonuniform energy distribution tends to reduce coastal irregularities. In bays, less energy is available per foot of shoreline, as indicated by the more widely spaced dashed lines, so erosion will be less there and sediment may be deposited.

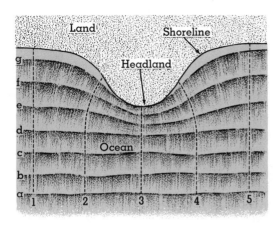

FIG. 12-6. **Wave retardation related to water depth.**

FEATURES OF GENTLY SLOPING BOTTOMS

Other processes are thought to occur on bottoms of low slope. As a wave approaches the shore, frictional forces are so distributed that the wave becomes asymmetrical and the crest moves landward faster than the rest of the wave. Water at the crest takes on a horizontal motion and the wave breaks, the water rushing intermittently up on the beach in a tumbling, turbulent motion and reaching some distance up the slope above general sea level. Water returns by flowing down the slope in a more nearly continuous manner and with less velocity.

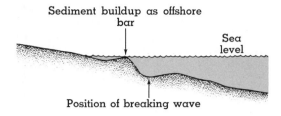

FIG. 12-7. Cross section of offshore bar formation (vertical exaggerated).

The breaker line (the line where waves break) is a place of high turbulence, and a trench is eroded on the bottom here (Fig. 12-7). The coarse sediment is carried landward by the more competent rushing water of the breaking waves, and only the finer sediment is moved seaward by the slowly returning flow of water. By this process sediment is shifted from place to place, and waves build up the beach under some conditions and erode it under others. Changes in sea level because of rising and falling tides, rising water levels occasioned by long-continued, strong onshore winds, and other changes resulting from waves of different heights and lengths cause the trench to shift position from time to time. It is believed that this process is responsible for formation of a submarine sand bar parallel to the coastline. The bar grows as sediment accumulates and may eventually be built up above sea level by large storm waves. Long, narrow islands, called *offshore bars*, or *barrier bars*, formed in this environment are to be found on many coasts of the world, notably the Gulf Coast of Texas (Padre Island) and Florida, and the Atlantic Coast of

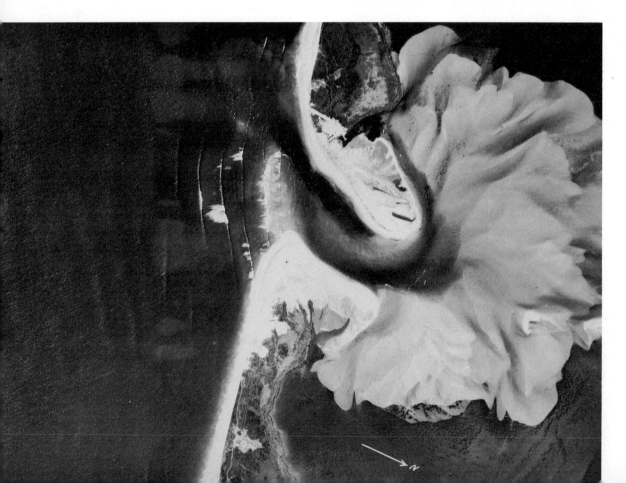

FIG. 12-8. Aerial view of Moriches Inlet, Long Island, New York. Waves from lower left move water and sediment from bottom to top of photograph, extending lower (eastern) island toward west and across inlet. Sediment also is transported to right (northward) and deposited behind island. (U.S. Air Corps.)

North America from North Carolina to Florida. A *lagoon*, ranging in width from a few feet to a few miles, exists between the offshore bar and the mainland. As time goes on, the lagoon tends to be destroyed by filling with sediment from the mainland, by shoreward transportation of sediment through inlets (Fig. 12–8), and by landward migration of the offshore bar as sediment is eroded from the seaward slope by storm waves, carried over the bar, and deposited on the lagoon side.

Along oceanic coasts lagoons are likely to contain water that is either more or less saline than normal sea water, depending on the relative importance of stream flow and evaporation. Few species of plants or animals can tolerate such unusual environmental conditions, but those that are present are likely to be abundant because of the lack of competition. Highly organic sediments, perhaps even peat, may be deposited under such conditions. The difference between such lagoonal sediments and the sand of the offshore bar is very marked, and the shoreward migration of sand of the bar, covering the organic sediment of the lagoon, leaves an unmistakable record observed in many places in the world.

On some coasts three distinct offshore bars may be found. Such an array is not readily explained by the mechanics described above, which suggests the possibility that some other sequence of steps may be responsible for the origin and development of offshore bars.

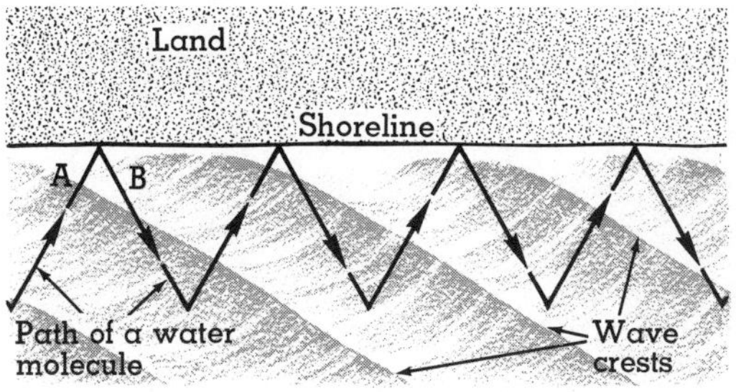

DEPOSITIONAL FEATURES FORMED
BY CURRENTS PARALLEL TO THE SHORELINE

The erosional and depositional forms discussed so far can result from wave action that impinges on the coast at any angle. But some depositional landforms are built only when waves approach the shore in such a way that the crests are not parallel to the shore line.

Waves moving toward a land area with crests oblique to the shore are shown in Fig. 12–9. Refraction in shoaling water causes a change to a pattern more nearly parallel, but complete parallelism is not achieved. Motion of a water molecule in the nearshore zone is shown in Fig. 12–10.

FIG. 12-10. Sediment transport in a longshore current.

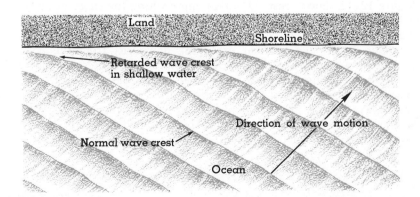

FIG. 12-9. Waves obliquely approaching shore.

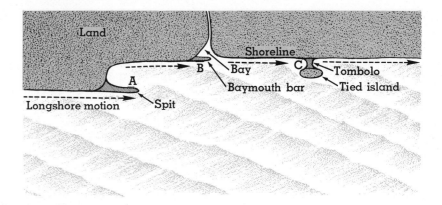

FIG. 12-11. Waves obliquely approaching shore originate longshore current and longshore drift (here moving from left to right), which deposit sediment to form spits, baymouth bars, and tombolo.

The molecule moves perpendicularly to the crest, as in path A. Motion is a combination of two directions, one parallel to and one perpendicular to the shoreline. After the wave breaks and water rushes up the slope, the water returns to the sea but continues its motion parallel to the shore, as in path B. Repetition of the two motions—first landward and seaward, and second parallel to the shore—is indicated by the zig-zag route of the arrows. The sum of the individual movements constitutes a current moving parallel to the shore that may include much water traveling great distances in a day. This is known as a *longshore current*. It has competence and capacity and may transport a large amount of sediment. The movement of sediment is called *longshore drift*. It can be measured by using tracers, and has been noted in some areas to have a velocity of ½ mile per day and to move many tons of clastic particles past a fixed point within a day.

Figure 12–11 shows crests of waves approaching the shore at an angle, and longshore current and longshore drift indicated by arrows. At point A the longshore drift will be deposited in deeper water and will eventually accumulate as an elongate bar extending downdrift from the point of land to form a *spit*. Waves approaching from another direction may alter the spit by recurving its end to make a *hook*. A similar deposit at point B is called a *baymouth bar*. An island at point C protects the shore behind it from wave action, and the longshore current in that area may be so weakened that sediment in transit may be deposited in the quiet zone, as shown in the figure. Such a deposit connecting an island to the mainland is called a *tombolo* (from an Italian word meaning pillow). Note that in Fig. 12–12 longshore drift transported sediment from the area of active erosion away from the observer and deposited it on the updrift (nearer) side of the pier.

SHORELINE HISTORIES

Relatively rapid rise or fall of water level with respect to the land (or the reverse) is not an uncommon phenomenon, on both oceans and lakes. Erosional and depositional features of shores may be submerged beneath a rising water level or may be lifted high above the water level far out of reach of the shoreline processes of aggradation and degradation. Indications that the land has been raised or the sea level lowered in an area are as follows:

FIG. 12-12. Cerro Azul, Peru. Sea cliff, at front. Material eroded from cliff deposited on near side of pier by longshore current. Note breakers showing waves approaching shoreline obliquely. (George R. Johnson.)

FIG. 12-13. Horizontal terrace above smelter marks high-level shoreline of a much larger lake of which Great Salt Lake, Utah, is a remnant. Black Rock, Utah. (The American Museum of Natural History, New York.)

1. Elevated sea cliffs and wave-cut and wave-built terraces (Fig. 12–13).
2. Elevated beach and deltaic deposits.
3. Gently sloping bottom and nearshore areas with offshore bars and lagoons.

Areas where the land has been submerged are characterized by these features:

1. Extremely irregular coastlines with many estuaries.
2. Drowned river valleys and fiords.
3. Wave-cut cliffs on headlands.
4. Spits, baymouth bars, and similar features formed by longshore drift.

Although many attempts have been made to classify coasts according to their histories, the combination of crustal mobility and numerous recent changes in sea level has caused most coasts to show a mixture of features of both the above categories.

SUGGESTED REFERENCES

Gibson, Count D.: *Sea Islands of Georgia*, University of Georgia Press, Athens, Ga., 1948. (Paperback.)

Guilcher, Andre: *Coastal and Submarine Morphology*, Methuen & Co., Ltd., London, 1958.

King, C. A. M.: *Beaches and Coasts*, Edward Arnold, London, 1959.

Leet, D. J., and Sheldon Judson: *Physical Geology*, 3rd ed., Prentice-Hall, Inc., Englewood Cliffs, N.J., 1965.

Rogers, John J. W., and John A. S. Adams: *Fundamentals of Geology*, Harper & Row, Publishers, Inc., New York, 1966.

Shelton, John S.: *Geology Illustrated*, W. H. Freeman & Co., San Francisco, 1966.

EARTH HISTORY

CHAPTER 13

Like every physical object, the earth has a history—it was formed, has undergone change, exists today, and will exist for a long time in the future. Like most things, it is acted upon by forces exterior to it. We have seen how the process of being illuminated by the sun affects the atmosphere and the oceans. Unlike most other objects with which we are familiar, the earth is so large and so complex that it acts upon itself, modifies environments, and causes changes at the surface and at depth. This has been illustrated by the interaction of the atmosphere, oceans, climate, vegetation, soil, and the erosional and gradational processes that have sculptured the face of the earth. We have some comprehension of at least the results of internally applied energy, if not the reasons for it or the ways it is applied, in what we have seen of vulcanism, igneous intrusion, deformation of the earth's crust, and the formation of igneous and metamorphic rocks. Notice that much of our knowledge of the processes affecting the earth has come from studying the evidence of its past. This suggests a good reason for at least a brief consideration of earth history. For our purposes, a discussion of the principles used in working out the history of the earth will be more useful than a recital of the events that have occurred. For those interested in the history itself, numerous books are available.

Aside from the purely intellectual quest for knowledge, why do humans inquire into the past of their planet? Even the old-time prospector, anxious to stake a claim and pan for gold, utilized without necessarily recognizing it the fruits of many investigations of earth history when he concentrated on stream and shoreline sand and gravel deposits. In a more sophisticated way, the modern exploration geologist analyzes ore deposits, determines the sequence of events responsible for their occurrence, and then seeks evidence of similar events in areas not yet productive of mineral wealth. The process is similar in the search for nonmetals, including diamonds at one end of the scale and such prosaic but essential commodities as salt, oil, and construction materials at the other extreme.

PRINCIPLES

Three simple but profound principles are used in determining the sequence of events in the history of the earth. The first is *uniformitarianism*, the principle that the events of the past were no different from events we see today. Chemical and physical laws are unchanging, and insofar as they govern geological processes the principle is valid. The biological environment has not been constant, for organisms have changed through evolution, and to this extent the events of the past are not repeated; in this respect the principle does not hold true. In addition, we find some ancient geologic materials that are unlike any deposits we observe being formed today; thus it is likely

FIG. 13-1. Cross-bedded sand exposed in channel eroded into a delta, Santa Cruz Irrigation Reservoir, New Mexico. Note how inclined laminations are cross-cut by the overlying horizontal layers. (U.S. Department of Agriculture.)

either that the full range of earth environments does not exist today or that we are not able to recognize or properly associate cause and effect in some instances. Much work is being done in this direction, and more will be necessary before we can answer fully the many questions that face us.

We apply the principle of uniformitarianism successfully when we are able to determine the past positions of land areas, shorelines, and ocean basins by studying the sedimentary rocks. In one area we may perhaps see a heterogeneous group of conglomerates, sandstones, and shales, in which cross-bedding shows frequent and extreme changes of current direction and velocity in bodies typical of stream channel deposits (Fig. 13–1: recently deposited deltaic sediment, as yet uncemented), alternating with dune sands (Fig. 13–2: cemented cross-bedded sandstone, perhaps 150 million years old). If the rocks were red and showed signs of weathering, and soil zones were recognizable, we would conclude that these sediments were deposited in a land area. Such structures as raindrop imprints and mudcracks preserved in the rocks would indicate occasional exposure to the air. Sediments deposited on the ocean bottom would have a different set of characteristics, and some of these would be preserved in the rocks.

The principle of *superposition* is another key concept. By this we mean that when sediments are deposited, the first (oldest) layer is beneath later (younger) layers. With this knowledge, we can establish a sequence of deposits and the corresponding relative dates on a time scale. Note that the actual years cannot be determined; but if the sequence is then tied at intervals to radioactive ages, an absolute time scale can be worked out. The major units and some specific ages are shown in Table 13–1, the

T A B L E 13 – 1   THE GEOLOGIC TIME SCALE

ERA	PERIOD	MILLIONS OF YEARS AGO	RADIOACTIVE DETERMINATIONS
Cenozoic	Quaternary		
		1	
	Tertiary		
		70	71 Boulder, Montana
Mesozoic	Cretaceous		
		135	
	Jurassic		
		180	
	Triassic		190 Palisades, New York
		225	
Paleozoic	Permian		
		270	
	Pennsylvanian		287 Kuttung, Australia
		300?	
	Mississippian		
		350	
	Devonian		
		400	390 East Greenland
	Silurian		
		440	
	Ordovician		
		500	500 Kolm, Sweden
	Cambrian		
		600	
Proterozoic			
		2500	
Archean			

FIG. 13-2. Cross-bedding in cemented sandstone, Oak Creek Canyon, Arizona. (U.S. Department of Agriculture.)

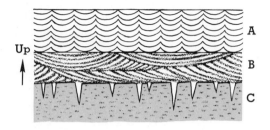

FIG. 13-3. Sedimentary structures used to determine top and bottom. A, ripple marks; B, cross-bedding; C, mud cracks (note filling by sediment like that of the overlying layer).

geologic time scale, which is represented as strata in sequence, the youngest at the top. The largest time units are the eras, which are divided into periods. These are generally accepted as representing about the same times at different parts of the earth. Smaller subdivisions of geologic time are not all as widely equivalent and are not given. Specific dates (in millions of years before the present) are given for a few radioactive determinations.

Bedded rocks can be overturned during folding, so it is essential to know whether each bed, especially where deformation of the crust has occurred, is in its original position or has been overturned, in order to determine the sequence of deposition. Some sedimentary structures can be used to answer this question. Oscillation ripple marks left on water-deposited sediment have sharp crests and gently rounded troughs (Figs. 13–3, bed A, and 13–4). In cross section in a rock they can be used to determine which way is up and therefore which bed is older. Lamination in units of cross-bedded sands and sandstones are crosscut at the top and are tangent to the bottom (Fig. 13–3, bed B). Mudcracks in finer

FIG. 13-4. Ripple marks on a bedding plane in sandstone. Compare with ripple marks on uncemented sand in Fig. 11-16. (American Geographical Society.)

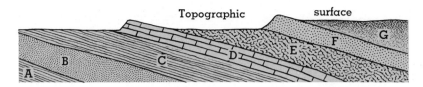

FIG. 13-5. Cross section of tilted sedimentary strata.

sediments may also be of use, since the cracks widen upward (Fig. 13–3, bed C). Many other features of inorganic and organic sedimentary rocks can be used to determine the original positions of the strata.

The final principle concerns *crosscutting*. It states that any structure which crosscuts a material body represents the later event. Examples are faults and dikes, which were formed after the rocks in which they occur. Another crosscutting relationship is the *unconformity*, which we shall discuss in some detail.

UNCONFORMITIES

Unconformities play a major role in enabling us to determine the sequence of events in the geologic history of a given area. An unconformity is a surface that separates younger rocks from older ones. We shall consider a number of examples.

The geologic cross section of Fig. 13–5 represents several events. If the rocks shown were deposited on the ocean bottom and have not been overturned, the following is assumed to have occurred:

1. Beds A through G were deposited below base level in alphabetical sequence, different lithologies being formed as the result of different materials and environments of deposition. The beds were originally horizontal.
2. The rocks were then tilted and subsequently raised above base level (order of these two steps not determinable here).
3. Erosion (J) removed the materials once present above the topographic surface. Because of the gentle dip of the beds, ridges were formed on the more resistant rocks (beds D and F). At the time of the profile, base level may not have been reached.

The solid earth being as mobile as we know it to be, we can assume that the materials were ultimately lowered until below sea level and therefore below base level (Fig. 13–6). Deposition of another horizontal sequence of

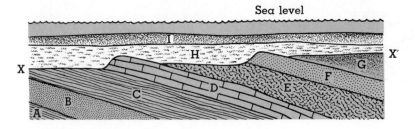

FIG. 13-6. Tilted and eroded sedimentary rocks submerged and covered by other sediments.

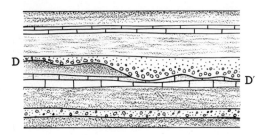

FIG. 13-7. Cross section of disconformity (D-D'). Sedimentary rock overlying sedimentary rock.

sedimentary rocks (H and I) occurred. Renewed uplift above base level without tilting, followed by erosion, exposed the rocks of Fig. 13–6 in cross section on the wall of a canyon. The heavy line X-X' is the unconformity, here the type known as an *angular unconformity*, because the beds above the unconformity are not parallel to those below. It marks the old erosion surface and represents a period of erosion. The contact of the bed just above the unconformity with the beds below it may be so distinct as to be a knife-edge in thickness and without any alteration of the underlying rocks, or there may be a weathered zone or perhaps a soil zone with the normal horizons.

If the sequence of steps lacks tilting and beds above and below the unconformity are parallel, we say there is a *disconformity* (Figs. 13–7 and 13–8). If the unconformity is underlain by rock in which tilting cannot be seen, such as granite, the noncommittal term *nonconformity* is used (Fig. 13–9). Erosion surfaces range from extensive plains with little relief on bare unaltered bedrock to maturely dissected surfaces with a thick soil overlying intensely weathered bedrock; and unconformities can be seen with the same range of conditions.

INTERPRETATION OF EVIDENCE IN THE ROCKS

Interpretation of past environments, events, and time is aided by the study of fossils in the rocks. A *fossil* is any recognizable organic structure or impression thereof preserved from the geologic past. Fossils range from the frozen and still edible complete bodies of wooly mammoths preserved during the most recent glaciation through bony skeletons, more or less articulated, to tiny fragments of shells or other hard parts, and tracks of animals or imprints of plants. From them

FIG. 13-8. Disconformity. Medium-gray sandstone overlying light-gray sandstone. Line marking disconformity inked for emphasis. (L. Ogden.)

and from evidence in the strata we attempt to reconstruct the population and environments represented by the rocks, using living relatives of the organisms for comparison. Because organisms have evolved, the further back we go in time the less certain our interpretations become. For example, we know that starfish are now exclusively marine in habitat. Presumably any rock in which we find fossils of starfish was deposited on an ocean bottom. But what of a shrimplike form? Some fresh-water animals are very similar to shrimp, so the salinity of water in which such crustaceans of the past lived cannot be estimated so reliably.

Broad interpretations may establish the positions of land areas and ocean basins; shorelines and to some extent particular depth zones in the sea may also be identified. The applications of such analysis are clearly seen in the search for petroleum, where we look for marine source beds adjacent to porous and permeable shoreline sand deposits that may be reservoirs in which oil and gas have accumulated. By studying vegetation and weathering products we can come to some conclusions about prehistoric climates.

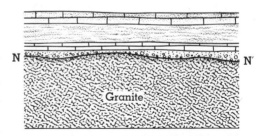

FIG. 13-9. Cross section of nonconformity (N-N'). Sedimentary rock overlying granite.

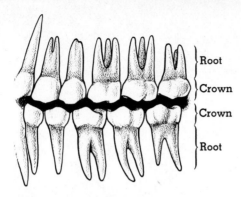

FIG. 13-10. Opposing teeth designed for crushing food between surfaces of the teeth.

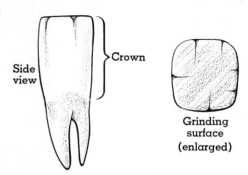

FIG. 13-11. High-crowned grinding tooth.

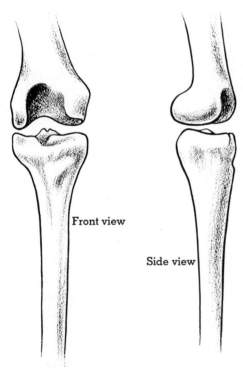

FIG. 13-12. Elongated leg bones. Joints have interlocking ridge and groove on cylindrical surfaces.

The form of skeletons may be useful in determining the functions of organisms, and further assumptions may be made from these. Specialized structures of animals are most helpful in this kind of inquiry. The teeth of some large mammals have comparatively short crowns bearing crushing surfaces, diagrammatically shown in Fig. 13–10. Other mammals are equipped with grinding teeth having much longer crowns and an intricately folded pattern of enamel (Fig. 13–11), which are worn down during the life of the animal. A conclusion confirmed by observation is that animals with crushing teeth browse on foliage and tender twigs growing some distance above the ground. In contrast, animals equipped with high-crowned, infolded-enamel teeth graze on grass growing close to the ground, with which sand and silt are ingested; the tooth form evolved because of the need for a long-lasting tooth capable of coping with appreciable amounts of abrasive sediment. Fossil teeth tell us not only about the animals but also about the vegetation, which, because of its delicacy, may not itself be represented in the fossil record.

Another specialized structure often studied is the limb of a vertebrate. What form of animal would have short, thick leg bones with large, flattened ball-and-socket joint surfaces? A reasonable conclusion would be that it was large, heavy, and slow-moving, the joints not permitting very much limb motion. In contrast, the long legs and joint design shown in Fig. 13–12 suggest an animal with restricted direction of movement and high-speed motion. Carried further, this suggests, as the physical environment, a plain or gently sloping land surface on which high speed is possible, and perhaps an associated predator that is also capable of fast motion. Other animal and plant structures can be interpreted and used as a basis for extrapolation in a similar manner.

EARTH SCIENCE

CORRELATION

By using the principles and methods discussed thus far, we may encounter considerable success in reconstructing the past positions of continents, oceans, and shorelines, but there is a further problem we have not yet considered. Let us assume we can see from the rocks that an area was at one time an ocean, represented by limestones containing fossils of marine plants and animals (Figs. 13–13, bed A, and 13–14); followed by uplift to become an arid land area, represented by a disconformity overlain by sandstone composed of cemented, cross-bedded dune sands (Fig. 13–13, bed B); then sinking and becoming a shallow ocean in which clay mud was deposited, since changed to a fossiliferous shale (Fig. 13–13, bed C). The same sequence may be found 200 miles away (Fig. 13–13, beds D, E, and F). Do bed A and bed D represent bottom sediments deposited during the same time period? Not necessarily! Were bed A and bed F deposited concurrently? Perhaps yes, perhaps no. In each case the environments were alike. What about bed A and bed E? Could they have been deposited simultaneously? This is possible—the area of point 1 could have been an ocean at the time point 2 was a continent, and a shoreline would have been between them.

Perhaps the problem is obvious to you now. We must establish the contemporaneity of events in different places to make comparisons of his-

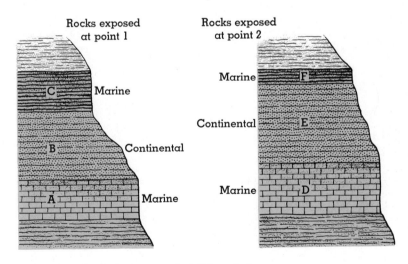

FIG. 13-13. Rock strata exposed at different places.

FIG. 13-14. Fossils on a bedding plane in a limestone.
(Ward's Natural Science Establishment, Inc.)

tories at these places. This process is called *correlation*. How can we correlate bed A and bed D? This may be accomplished in several ways. Where soil is thin and bedrock is widely exposed, as in desert or mountainous areas, it may be possible to walk along a continuous or nearly continuous outcrop for many miles. The rimrock of the Grand Canyon of the Colorado is an example. If the horizontal distance is not great or the bed represents simultaneous deposition—as, for example, a layer of volcanic ash an inch or so thick—we can be certain of correlation. However, over great distances even a single bed does not represent exact correlation. A sandstone may mark nearshore deposition of sand by an ocean with constantly rising surface, and as the shoreline migrates landward it leaves behind it a sheet of sand older in the seaward direction. If the exposures are not continuous but similar sequences and rocks are found at different places, we can correlate the like beds with each other. The shorter the distance and the greater the similarities, the more certain the correlation. Fossils may play a key role in correlation. Some forms existed for only short times but spread over great distances, either as living organisms or perhaps transported after death by the wind, waves, or currents. When buried in sediment they became permanent markers for that particular time, even though the sediments in which they were deposited and now the rocks we find them in are different. These forms are called *index fossils*, and they are particularly valuable in correlating across areas without rock outcrops or from one continent to another. With them it is possible to establish exact dates of deposition of a bed of rock at a given location by using radioactive minerals and fixing the same date in the sequences of rock in other areas of the world.

SUMMARY

By field study of the rocks, laboratory investigations of samples, organization of the information obtained into a three-dimensional whole, and, finally, relating the materials and their present distribution to the fourth dimension, time, we try to construct the geologic history of a particular point, region, continent, or the world. Under the best conditions, we may produce a surprisingly detailed picture of a complex sequence of events. The process requires the fullest application of the scientific method—including the collection of data, their organization into a rational system, preparation of a hypothesis that takes into account every detail, testing it by making predictions as to yet undiscovered phenomena, and finding them in the field; and, as so often happens, turning up new evidence requiring revision of the hypothesis and repetition of the process.

In almost all investigations of earth history, there are two great challenges: First, no single field of knowledge is adequate to solve a given problem—the basic sciences plus mathematics, engineering, and geology are all involved; and, second, only incomplete evidence is available with which the problem can be attacked. Some earth scientists meet the first challenge by studying several fields in depth. Many become specialists in one field and acquire a general familiarity with the rest; others become members of a team, each working on his own part of the problem, with frequent interchange of ideas, reports of progress, and requests for help.

The second challenge may be less satisfactorily solved. Often it appears that there is more than one possible geological route to the final product. How can one look at the number 16 and tell whether it came into being by adding 2 and 14, by adding 9 and 7, or by dividing 528 by 33? The earth scientist attempts to meet this difficulty by setting up multiple working hypotheses and investigating each possible solution until one becomes established, much as a police detective checks out more than one suspect in a murder case.

SUGGESTED REFERENCES

Eardley, A. J.: *General College Geology*, Harper & Row, Publishers, Inc., New York, 1965.

Rapport, Samuel, and Helen Wright (eds.): *Crust of the Earth*, New American Library of World Literature, Inc., New York, 1955. (Paperback.)

Stokes, William L.: *Essentials of Earth History*, 2nd ed., Prentice-Hall, Inc., Englewood Cliffs, N.J., 1966.

Woodford, A. C.: *Historical Geology*, W. H. Freeman & Co., San Francisco, 1965.

WORLD PATTERNS

CHAPTER 14

14

The processes that form the natural environments of the earth have been discussed individually in earlier chapters, with emphasis on the elements comprising each type of occurrence. We have seen, however, that the natural phenomena of our planet exist not independently but in important interlocking relationships. Thus the distribution of land masses and the elevations of land surfaces affect the extent and boundaries of the various climatic zones; climate in turn controls natural vegetation; and these two together largely determine the different types of soil, which delimit the nature of the agricultural production that can be successfully practiced in any locality. The existence of mineral deposits depends not on present conditions but on the climatic events of the past.

In this chapter we shall return to a number of the most important aspects of the earth's natural phenomena previously considered, with emphasis now on (1) the correlations and connections between them, (2) the patterns of their distribution over the globe, and (3) the relationship of these patterns to man and modern society. The worldwide patterns of distribution are of special significance because of the light they shed on the great disparities in the modes of life prevailing in the various regions of the earth, and on the fact that an overwhelming proportion of the earth's people occupies a very small percentage of its surface. As a final topic, we shall give some consideration to the distribution of man over the area of the world, and to his use of and impact on the physical environment.

NATURAL VEGETATION

The natural vegetative cover of the earth's localities continues as an important factor in our lives, despite the growing predominance of the technological aspects of society. The forests provide lumber, pulp, fuel, foods, and medicines; the grasslands serve as natural pastures for livestock. Vegetative cover is an important link in the hydrologic cycle as a retainer and regulator of water; it is a habitat for wildlife, and a vital and refreshing recreational resource for man.

The natural vegetative cover of a locality depends on local and regional climatic, topographic, soil, and biological conditions. Plants are very closely attuned to climate, for they are exposed to the elements with little means of protection, and none of escape. Specifically, they are sensitive to these climatic factors:

1. Temperature—Each species has maximum and minimum limits beyond which it cannot exist, and an optimum for maximum growth.
2. Moisture—Certain species are especially adapted to thrive under extreme conditions; some, known as *xerophytes*, have the ability to

withstand drought; others, the *hydrophytes*, are adapted to extremely wet conditions.

3. Length of frost-free season.
4. Duration of daylight.
5. Angle of the sun.
6. Wind velocity.

Vegetation is also influenced by local variations in topography and soil: drainage, slope, type of bedrock, texture and structure of soil; and by such biological factors as the type of insects and viruses.

Certain plant species tend to appear together in a particular type of environment, such as a pond or a rock cliff. Such groups of species are known as plant associations. Broader groupings adapted to the major climates are called plant formations. Desert shrub, for example, is very similar in general appearance in the *BWh* climates of the Sahara, northern Mexico, and central Australia, even though individual species may differ in particulars.

Speaking on a worldwide scale, natural vegetation may be divided into four broad categories: (1) tundra, (2) desert shrub, (3) grasslands, and (4) forestlands. Note that the first three are plant formations but that the last is an even larger grouping, which is further subdivided.

TUNDRA

Tundra comprises lichens, mosses, sedges, hardy grasses, and stunted trees—all xerophytic (Fig. 14–1). In areas where these plants appear, soil

FIG. 14-1. The Arctic tundra. The tundra is a treeless expanse consisting of a variety of hardy grasses, mosses, sedges, and lichens. The tundra landscape shown is the grass or meadow tundra, which may be used as pastureland for reindeer or caribou. (Richard Finnie.)

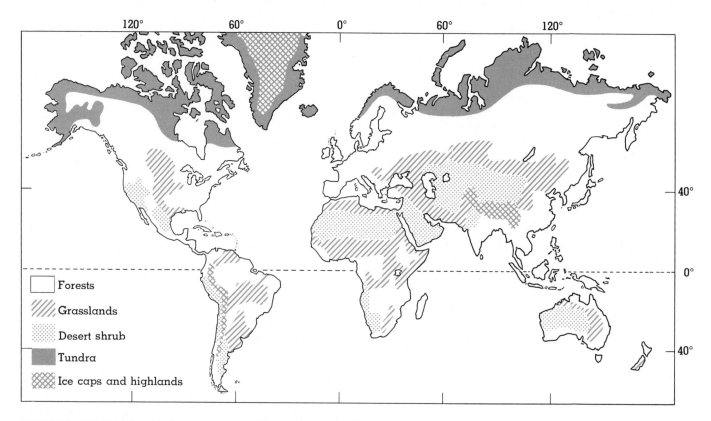

FIG. 14-2. Highly generalized
map of the major plant groups.

Forests

Grasslands

Desert shrub

Tundra

Ice caps and highlands

moisture is in the form of ice or snow most of the year and is thus not available for plant use.

Tundras are found in *ET* climates and are located beyond the tree line (Fig. 14–2). They differ in their specific composition from one location to another. Along the equatorward margins of climatic regions where they are found an abundance of dwarfed trees is in evidence; in swampy areas mosses and muskegs are most common; and in drier regions lichens, sedges, and grasses predominate.

Perhaps as much as 3 per cent of the land area of the earth is covered by tundra vegetation.

DESERT SHRUB

The *BW* climates are associated with desert shrub, a sparse vegetative cover adjusted to severe dry conditions. Plants adapt to prolonged drought in two major ways: They either evade dry conditions or they resist drought periods. Most annual species are able to evade excessively dry periods because of their very short life cycle. Seeds remain dormant during prolonged dry periods, germinating only after a rain and then growing rapidly to maturity within a 5- to 6-week period. The perennial species resist droughts—by shedding leaves during excessively dry periods, by storing water in fleshy parts of the plant or developing long tap roots, or by the modification of leaves to spines thus reducing the transpiration rate.

Desert shrub includes hardy bunch grasses, cacti, and perennial shrubs (Fig. 14–3). About 15 per cent of the land area of the earth is covered by this type of vegetation.

FIG. 14-3. Desert shrub. Desert shrub includes a variety of cacti, hardy grasses, and other types of xerophytic vegetation. This type of vegetative cover is of little value for grazing. (U.S. Department of Agriculture.)

GRASSLANDS

About one-third of the world's land area is covered by natural and improved grasses. Generally, grass associations predominate where rainfall is not sufficient for tree growth, in areas having more moisture than those where desert shrub prevails, and in areas with longer frost-free seasons than the tundras. Grasslands include

tropical savanna grasses,
prairie grasses of the midlatitudes, and
the short grasses of the tropical and midlatitude steppes.

Tropical savanna grasses range from 5 to 15 feet or more in height and are usually found in *Aw* and *As* climates, where a long dry season imposes a dormant period (Fig. 14–4). They also prevail in tropical areas of impermeable soil and poor drainage, and those where man periodically burns off the vegetative cover to expand the grassland area.

Prairie grasses are usually 2 to 8 feet in height and are found in the subhumid sections of the humid subtropical and humid continental climates.

Steppe grasses are short, sparse grasses, 1 to 2 feet in height; they are associated with the *BS* climates (Fig. 14–5). They are used to some degree as pasturelands and in certain areas have been cleared for the production of wheat or other drought-resistant grains.

In North America, much of the original prairie grass has been cleared to make room for agricultural crops. Perhaps one-half the total grassland area is used for pasturing livestock.

FIG. 14-4. Savanna, or tropical grasslands. Note widely spaced, small semideciduous trees adapted to long dry periods. (**U.S. Department of Agriculture.**)

FIG. 14-5. Steppe grass, typical of the transition zone between desert and humid climates. Although bunched or sparse in character, the steppe grass has value as low-carrying pasture for livestock. (U.S. Department of Agriculture.)

FIG. 14-6. Selva, or tropical rainforest. Note the stratification of trees and the great variety of species. (Alexander Hamilton Rice.)

FORESTLANDS

The forests once covered about one-half the land area of the earth, but they have been cleared from many areas that were considered suitable for pasture or cropland. Presently, they occupy about 30 per cent of the total land area. They may be roughly grouped into tropical rainforests, lighter tropical forests, Mediterranean woodlands, midlatitude broadleaf forests, and coniferous forests.

Tropical rainforests are unique in that several hundreds of species of trees may be found within a square mile. The continually warm and humid climate in which they are found allows vegetation to grow throughout the year without a dormant period. True tropical rainforests, referred to as *selva*, are evergreen and broadleaf. They are vertically stratified, or multistoried, and as many as three distinct layers or tiers may be distinguished (Fig. 14–6). An abundance of lianas, other climbers, and epiphytes, or air-root plants, are associated with the tropical rainforest complex.

Tropical rainforests make up about 47 per cent of the world's forest-lands and are associated primarily with the *Af* and wetter parts of the *Am* climates. Although they include many important commercial woods such as mahogany, balsa, dyewoods, natural rubber, and Brazil nuts, the selva forests are not widely utilized because of the lack of pure stands of commercial species, their great distance from the midlatitude markets, and the absence locally of labor, sanitation, and knowledge of modern logging techniques.

Lighter tropical forests comprise a variety of thorn, scrub, and deciduous and semideciduous trees. These forests are composed of smaller, more widely spaced, and more deciduous trees than the tropical rainforests, and they are characterized by a dense undergrowth of tropical grasses and shrubs. They are most common in the *Am* and wetter parts of the *Aw* climates but may also be found in *Af* climates under unusual soil or drainage conditions.

Mediterranean woodlands are associated with *Cs* climates and account for about 2 per cent of the forests of the earth. They are unusual in that the trees are evergreen but are also adapted to summer drought conditions. The Mediterranean woodlands are composed of dwarfed trees and shrubs, widely spaced with a parklike appearance (Fig. 14–7). They are referred to as *sclerophyll* woodlands. In California the shrubs, bushes, and stunted trees are known as *chaparral*. The olive, eucalyptus, and cork trees are the best-known commercial species.

Midlatitude broadleaf forests are widely divergent in composition; the dominant species of trees differ from one region to another depending upon climatic and soil conditions. The broadleaf forests are usually deciduous, shedding their leaves in the winter season (Fig. 14–8). However, along their subtropical margins evergreen broadleaf associations may be found. The broadleaf forests are most commonly associated with the wetter portions of the *Ca*, *Cb*, *Da*, and *Db* climates. In most regions, the original broadleaf forests have been cleared to make way for crop-lands or pastures. Remaining stands are generally found only in hilly sections or areas of poor soil.

Coniferous, or cone-bearing, forests account for one-half the total forest cover of the earth. They are found on the cold, dry, or windy margins of the broadleaf forests. Coniferous trees are predominantly evergreen, and the shedding of the needlelike leaves is not confined to a specific season. These species are somewhat xerophytic; they are adapted to extreme cold, dry conditions, and continuous strong winds.

The coniferous forests of the subarctic climates (*Dc, Dd*) are referred

FIG. 14-7. Mediterranean woodland, evergreen shrub usually consisting of dwarf, live oak, cork oak, olive trees and shrub. (U.S. Forest Service.)

FIG. 14-8. Broadleaf forest. The broadleaf forests of the midlatitudes usually consist of oaks, maples, hickory, and other broadleaf deciduous species. Much of the original forest land has been cleared for agricultural or pastoral purposes. (L. J. Prater.)

FIG. 14-9. Arctic coniferous forest. The taiga of North America is largely composed of spruces, pines, firs, and some aspen. (H. M. Raup.)

to as *taiga* and are composed primarily of larches, spruces, and pines (Fig. 14-9). Taiga forests are found in North America and Eurasia. The Eurasian taiga is the largest contiguous forest area in the world, stretching west to east from Pacific to Atlantic coasts. The poleward margins are bounded by treeless tundra, and along their southern borders they gradually merge into broadleaf forests.

Coniferous forests are also found in *C* and *Da* and *Db* climates in highlands or rugged areas, and in some coastal plains where sandy soils predominate, as in the southeastern United States. The porous, sandy soils create a condition of rapid drainage that is generally not suited to broadleaf species. The coniferous forests are a valuable source of lumber, pulp, and naval stores.

SOILS

Soil sustains human life, for directly or indirectly it provides most of what man consumes as food. Soils differ greatly in productivity from one area to another and thus largely determine the agricultural potentialities of a locality. Significant properties of the soil are (1) fertility, (2) texture, (3) structure, (4) organic components, (5) color, and (6) profile.

Soil fertility depends essentially on chemical composition. Fertile soils are those that have substantial supplies of available nitrogen, calcium, potash, and phosphorus, in addition to other elements needed in small amounts such as iron, sodium, magnesium, copper, zinc, boron, iodine, and manganese. The supply of vital minerals may be reduced by solution in ground water and transportation into the subsoil (leaching). Erosion and excessive cropping may also reduce fertility. The rate of leaching is largely determined by the amount of rainfall, hence humid regions generally have less-fertile and more-acid soils than dry areas.

Soil texture refers to the size of the soil particles. This is an important factor because particle size influences the drainage of ground water and the movement of air. Soils composed of large particles do not retain water and are thus not well suited to the growth of agricultural crops; soils having extremely small particles may inhibit the movement of water and air to such an extent as to be unproductive. The texture of soil varies from sands to silts to clays (the finest).

The structure of the soil depends on the way in which the individual particles are grouped together. In a well-structured soil, particles are in small clumps, which allow easy penetration of water and air.

The organic matter, or *humus*, in the soil provides food for beneficial microorganisms and for plants, helps in the retention of water, and promotes good structure.

The color of soil reflects its physical and chemical characteristics. In general, dark-colored soils have a high organic content, whereas whitish soils contain an excessive concentration of soluble salts and minerals.

The soil profile defines the characteristics of the soil's horizontal layers, or *horizons*, which differ from each other in physical and chemical properties. For all soils the A horizon is defined as the zone of organic accumulation and maximum leaching and eluviation (Fig. 14–10); B horizon is the zone of redeposition and illuviation; and C horizon is the slightly altered and weathered parent material. The various types of soil differ considerably in the thickness and specific composition of each horizon, as well as the extent to which there are clearly defined horizons.

Soils are classified in three large groupings called *orders*: zonal, intrazonal, and azonal. Zonal soils are mature and have well-developed profiles. They correlate closely in areal extent with the major climates and the major plant groups of the world. Intrazonal soils also have developed profiles but these profiles are the result of local conditions, such as drainage. Azonal soils have little or no profile development and may be found in any area regardless of climate or vegetation. Examples are the alluvial soils associated with deltas and floodplains; sand dunes; and volcanic materials.

The zonal soils may be grouped into eight broad categories: (1) lateritic soils, (2) podsols, (3) gray-brown podsolic soils, (4) prairie soils, (5) chernozems, (6) chestnut soils, (7) desert soils, and (8) tundra soils. In

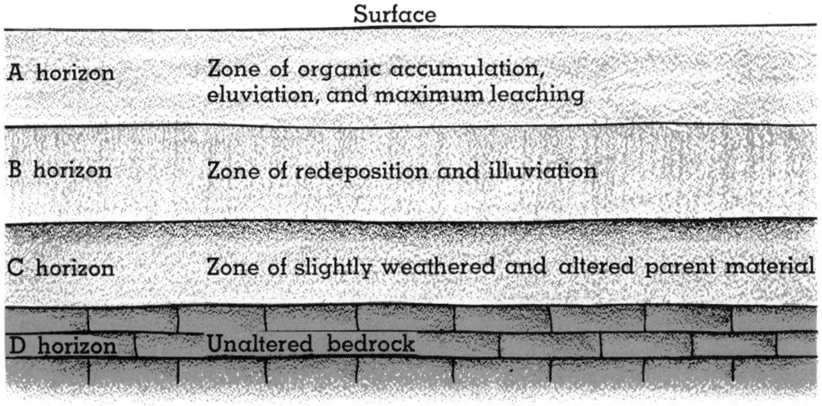

Surface

A horizon	Zone of organic accumulation, eluviation, and maximum leaching
B horizon	Zone of redeposition and illuviation
C horizon	Zone of slightly weathered and altered parent material
D horizon	Unaltered bedrock

FIG. 14-10. The horizons of a mature soil. Each horizon may be further subdivided on the basis of the speed and degree with which the soil-forming processes act, as well as other features.

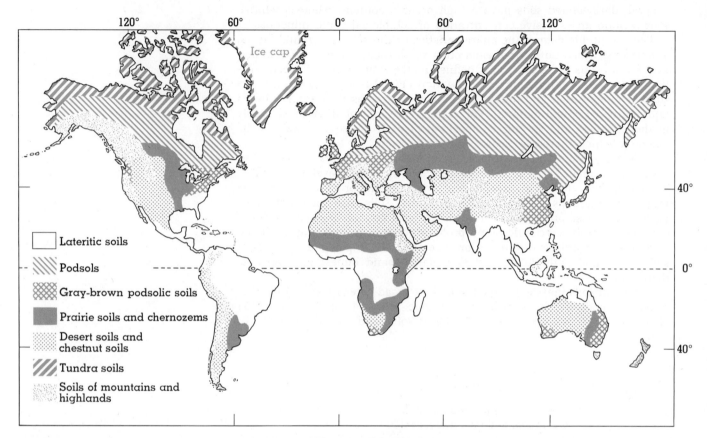

FIG. 14-11. Highly generalized map of the zonal soils of the world.

addition, mountain or highland soils cover a significant portion of the world's land and are usually included in the major soil groups.

Figure 14–11 shows the geographical distribution of the zonal soils.

LATERITIC SOILS

These soils are characteristic of tropical and subtropical humid climates. They are red and yellow, and they develop to considerable depths. In most locations *A* horizon is deep, coarse-textured, and porous. It is usually deficient in organic matter and essential plant foods owing to excessive eluviation and leaching associated with tropical temperatures and heavy rainfall. The *B* horizon is a darker or deeper red than *A* and is relatively fine textured. The infertility of lateritic soils does not permit continuous cultivation without heavy fertilization. This group is found in the *A* climates and in the wetter and more subtropical parts of *C* climates.

PODSOLS

These soils normally develop under coniferous vegetative cover in sub-arctic climates. A layer of needles from the coniferous vegetation usually covers the ground and adds to the acidity of downward-moving water. This results in an excessively leached, gray-to-white *A* horizon. *B* horizon is heavily illuviated and may develop into a layer of stony material called *hardpan*. Podsols are shallow soils, usually less than 2 feet in depth, and are low in essential plant foods. They are not normally used for agricultural purposes unless improved by heavy fertilization. The podsols predominate in the taiga forests of North America and Eurasia and in the cooler forest areas of the humid continental, short-summer climate.

GRAY—BROWN PODSOLIC SOILS

These soils are characteristically found in humid continental climate regions covered by broadleaf deciduous forests. They are equatorward of the true podsols. Usually an accumulation of tree leaves covers the surface. The broad leaves contain more calcium, potash, and other minerals than the needleleaf covering of the podsols. As a result, *A* horizon is less leached than in the podsols and is gray to brown rather than whitish. Gray-brown podsolic soils are of medium fertility. They contain more organic matter and are better structured and less leached than podsols or lateritic soils. They are the best of the forest soils and are widely used for agricultural purposes. Under careful management they respond to fertilization, and crop yields are relatively high.

PRAIRIE SOILS

These soils develop under a tall prairie grass cover in temperate humid climates, particularly the drier parts of the humid continental and humid subtropical climates. The luxuriant grass cover results in a dark *A* horizon containing considerable organic matter, and the modest amount of rainfall results in little leaching. These dark-gray-to-black soils, high in available organic matter and mineral plant foods, are among the most productive zonal soils. The rich Corn Belt farmland of the American Midwest lies largely within the area of prairie soils.

CHERNOZEMS

These soils are found along the dry margins of the prairie soils. They develop under a thick mat of grass roots and are very high in organic material. The small amount of rainfall results in little or no leaching, and the chernozems are particularly fertile. *A* horizon is black and 2–3 feet thick; it grades into a gray *B* horizon containing whitish nodules of accumulated calcium minerals. Chernozems develop along the margins of the *C* or *D* climates and the *BS* climates under a short prairie or steppe grass cover. Agricultural yields are generally very high. Where yields are average or low, this is the result of meager precipitation rather than any inherent lack of fertility of the soil.

CHESTNUT AND BROWN SOILS

These soils are found on the drier margins of the chernozems in *BS* or semiarid climates. They develop under sparse grass cover and as a result contain less organic matter and are lighter in color than the chernozems. The semiarid climate, with little rainfall and high evaporation rates, causes an accumulation of lime and other alkaline materials near the surface. These brownish soils are well adapted to cultivation if irrigated. Normally, however, they are used for livestock grazing rather than the production of crops, because of the sparsity of precipitation.

DESERT SOILS

These soils are light in color, thin, and very low in organic matter. They develop in *BW* climates under desert shrub vegetation. Although low in organic matter and nitrogen, they usually contain large amounts of soluble minerals brought to the surface by capillary action of ground water. When irrigated they may prove to be productive if they are not excessively alkaline.

TUNDRA SOILS

These soils are found in the treeless tundra (*ET*) climates. They develop under excessive moisture due to the very low rate of evaporation and the permanently frozen subsoil. A surface layer of brown, peaty, partially decayed vegetation is underlain by a grayish horizon that usually is saturated with moisture. Below this layer, which often has a plastic or fluid character, there is usually a permanently frozen layer. A large part of the tundra has poor drainage, and this results in swampland and bogs. The severe polar climate prohibits any agricultural use of the land, but forage for reindeer and caribou herds is available.

MOUNTAIN OR HIGHLAND SOILS

Although mountain soils are not zonal, they cover a significant part of the world's land masses and for this reason may be included with the

FIG. 14-12. Gully erosion. Disastrous effect resulting from erosion caused by cultivating the land parallel to the slope. (H. H. Bennett.)

major soil groups. Because slopes, highland climates, and vegetation may be widely divergent from one mountain area to another, it is difficult to generalize about these soils. It is safe to say, however, that they are normally very thin, owing to excessive erosion, and are not widely used. The underlying parent material and slope are more significant in their development than climate and vegetation.

Many intrazonal or azonal soils are productive, and they may be important resources in a given locality. Alluvial soils of the deltas and floodplains, if fine textured, may be highly productive. Soils developing from loessial materials, volcanic ash, and glacial till are also often highly fertile.

Regardless of their natural fertility, soils can deteriorate rapidly under poor management (Figs. 14–12 and 14–13). The removal of vegetation followed by cultivation leads to accelerated erosion of fine-textured soil on steep slopes. Overgrazing or excessive cropping without fertilization can result in exhaustion of the soil. Measures that can be profitably employed to conserve the soil and sustain its productivity are application of fertilizers, contour plowing, strip farming, proper cropping and rotation practices, damming of gullies, terracing, and the return of high-slope lands and excessively eroded lands to permanent forest or grass.

FIG. 14-13. Sheet erosion. This land should probably not have been plowed because of the sandy nature of the soil. In addition to sheet erosion, note the incipient gullies. (Soil Conservation Service.)

EARTH SCIENCE

LANDFORMS

As discussed in earlier chapters, the face of the land varies greatly in slope, local relief, and profile. On the basis of these characteristics, landforms may be grouped into four broad classes: (1) plains, (2) plateaus, (3) mountains, and (4) hills. Their geographical distribution is shown in Fig. 14–14.

PLAINS

Plains are areas of relatively low elevation; they have a predominance of gently sloping land with low local relief. In general, the local relief is less than 500 feet, and in many plain areas it does not exceed 50 feet. The gentle slope of the land permits the development of agriculture, lines of transportation, and settlements, with little or no restriction. Where

FIG. 14-14. Highly generalized map showing the major landform regions of the world.

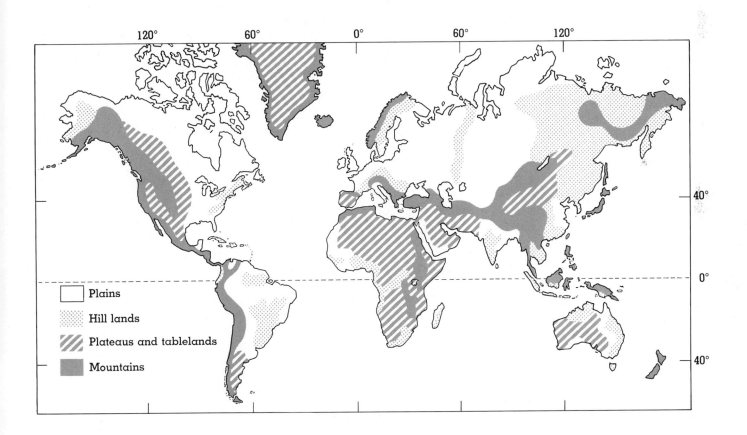

Plains

Hill lands

Plateaus and tablelands

Mountains

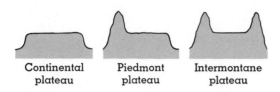

FIG. 14-15. Major types of plateaus. Continental plateaus are surrounded by lower lands; piedmont plateaus are bordered on one side by higher lands; and intermontane plateaus are surrounded by mountains or higher lands.

climate and soil are favorable, 90 per cent or more of the total area may be arable. Plains are more conducive to supporting large populations than any other class of landform; hence about 95 per cent of the world's population resides in these areas.

About 47 per cent of the land area of the world may be classified as plain lands. They may be deltas, floodplains, glacial till or outwash plains, lacustrine plains (former lake beds), uplifted coastal plains, stream-eroded flat lands, or karst plains. Extensive plain areas are found in eastern North America; the Amazon, Orinoco, and Parana-Paraguay regions of South America; central and western Eurasia; the Congo region and western Africa; and east Australia.

PLATEAUS

Plateaus, or tablelands, are areas of higher elevation than plains. Local relief ranges from 500 to several thousand feet. Like plains, they have a high percentage of level to gently sloping land, but they are dissected by deep river valleys or canyons often thousands of feet in depth. They may be classified as continental, intermontane, and piedmont, on the basis of their adjacent land areas (Fig. 14–15). They make up 5 per cent of the total ice-free land area of the earth. The ice caps of Antarctica and Greenland are continental plateaus and cover an additional 11 per cent of the earth's solid surface. Plateaus are created by uplifts of land. They are usually preserved from rapid stream dissection by a combination of such factors as their flatness, a resistant capping layer of rock, arid or semiarid climate, porous surface materials, and thick vegetative cover.

The tablelands of the midlatitudes are not used extensively for agricultural purposes as their high elevations usually result in cool temperatures and sparse rainfall. The deep canyons are barriers to transportation and communication and tend to isolate the settled areas from one another. In tropical areas, however, plateaus are often densely populated, since the high altitudes offset the usual high temperatures and humidity of the low latitudes. In high latitudes the plateau areas are excessively cold, and the ice plateaus of Antarctica and Greenland are the least desirable environments for human society.

The major plateaus of North America are the Colorado Plateau; most of the Great Basin between the Rocky Mountains and the Sierra-Cascades; the Columbia Plateau of northwestern United States; and the intermontane plateau, or mesa, of northern and central Mexico. The Brazilian highlands, Patagonia, and the high Altiplano, in the Andes, are the major plateau areas of South America. Over one-half the continent of Africa may be classified as a continental plateau; and the Iberian, Iranian, Tibetan, Arabian and Anatolian plateaus, along with the Massif Central of France, comprise the major tableland areas of Eurasia. The western half of Australia is low plateau, and the ice caps of Antarctica and Greenland are the major plateaus in the polar latitudes.

MOUNTAINS

Mountain areas are characterized by high elevations, great relief, and a very small amount of level or low-slope land. Local relief is at least 2000 feet and may exceed 10,000 feet in very high mountains. Generally, less than 1 per cent of the total land surface may be classified as low-slope land. About 27 per cent of the land area of the earth is mountainous, but less than 1 per cent of the world's population lives permanently in mountain areas. The steep slopes and low temperatures limit agricultural development to a few favored valleys and basins. The rugged relief serves as an effective barrier to transportation, communication, and cultural diffusion.

Mountains are the result of folding, faulting, and vulcanism, both extrusive and intrusive, and reflect unstable regions of the earth. The changes in weather and climate with increasing elevation have been discussed earlier (Chapter 6). Most of the great mountain regions are found in two major belts. One forms a ring of highlands around the Pacific Ocean and includes the Cascades, Sierra Nevadas, and Rocky and Aleutian Mountains in North America; the highlands of Central America; the Andes Mountains of South America; and the mountains of eastern Australia, New Zealand, New Guinea, and Japan. The second belt consists of three major extensions radiating from the Pamir Knot in southern Eurasia. One extension, trending east to west into Europe and North Africa, comprises the Hindu Kush, Elburz, Caucasus, Alps, and the Atlas ranges. A second extension, trending northeast, includes the Tien Shan, Altai, Sayan, Yablonovy, and Stanovoy Mountains; and the third arm extends east and southeast and includes the Karakorum, Kun Lun, Astin and Himalaya Mountains.

HILLS

Hills are characterized by moderate-to-high local relief ranging from 500 to 2000 feet. Usually, 5 to 10 per cent of such a region is suitable for agricultural purposes from the standpoint of slope. Hills may be the result of folding, faulting, extrusive or intrusive vulcanism, stream dissection of plateaus or high plains, glacial deposition such as a belt of moraines, or deposition of wind-blown sand.

These landforms cover 10 per cent of the land area of the world and appear in every continent. Extensive hill regions in North America are found in New England, the Laurentian uplands, the Appalachian region, the Ozark uplands, the foothills of the Rocky Mountains, and the coastal ranges along the west coast. Almost 15 per cent of North America is in the category of hills, no other continent having such a high proportion of its total area in this landform classification.

In South America, dissected parts of the Brazilian highlands and the foothills of the Andes Mountains represent the major hill areas. The Balkans, Scandinavian highlands, the Urals, most of Scotland, Wales, and central England, and much of Italy and Portugal make up the major hill areas of Europe. In Asia, most of northern China and adjacent areas are hills, as are central New Zealand, eastern Australia, central New Guinea, Borneo, and Java in Oceania.

MINERAL RESOURCES

As knowledge, culture, and technology have progressed, modern man has come to lean more heavily on the availability of mineral resources to maintain and improve his standard of living. The nations of the world can be ranked with respect to possession of mineral resources in four classes: (1) those which have an oversupply and are consequently exporters, (2) those with a balance of production and consumption, (3) those which must import minerals that are no longer available domestically in adequate quantities, and (4) those which must import many minerals, at high cost, because of the lack of domestic sources. It should be noted, however, that no nation is self-sufficient in all minerals, so that any nation would be in different positions with respect to different substances.

We shall examine briefly a few of the mineral resources used in the greatest quantities in the world, and their distribution.

ENERGY SOURCES

Most of the energy produced from mineral materials to the present time has come from the great deposits and reservoirs of coal, petroleum, and natural gas. Solar energy was slowly stored in organic matter buried hundreds of millions of years ago, and it is now being rapidly released by man to power his machines and to a lesser extent to provide the raw materials from which synthetic or substitute materials are made.

Coal is formed by the accumulation of nearly pure plant matter without deposition of clastic or chemical sediment. Modern swamps are thought to represent conditions of the past when the great coal beds were laid down. In such an environment as the Dismal Swamp of North Carolina and Virginia, a thickness of many feet of plant material has accumulated under conditions unfavorable to its destruction by organic or inorganic means. Burial results in compression and dewatering to form peat. Under the application of heat and continued pressure, volatile compounds are driven off, leaving coal of the various ranks, depending on the degree to which the metamorphic process continues. Up to a point in this sequence the amount of heat available per pound of coal increases, as does the desirability as a fuel; but before this stage is reached the coal passes through a period when it is quite valuable as a source of coke and volatile compounds from which a variety of useful products can be synthesized. A good grade of *coking coal* is essential in making the coke required for the production of pig iron.

Several million, million tons of coal are estimated to exist in deposits throughout the world. Although the total quantity is great, this resource is unevenly distributed among the nations and the continents. Table 14–1 shows the distribution of coal among the countries in which major deposits exist.

T A B L E 1 4 – 1   COAL PRODUCTION AND ESTIMATED RESERVES, 1963

COUNTRY	PER CENT OF WORLD PRODUCTION	PER CENT OF WORLD RESERVES
U.S.A.	22	40
China	22	20
U.S.S.R.	20	24
United Kingdom	10	4
Western Germany	7	7
Poland	6	1

In North America, anthracite coal in commercially valuable amounts is restricted to a small area in the Appalachian Mountains where the beds were much folded and metamorphosed. Little anthracite is mined today. Bituminous coal is present in extensive fields along the western part of the Appalachian Mountains, in large areas of Indiana, Illinois, Kentucky, Iowa, Missouri, Kansas, and Oklahoma, and in many small areas within the Rocky Mountains. Large deposits of low-rank bituminous coal and lignite occur in the northern Rocky Mountains and the High Plains States (Montana and North Dakota) and Provinces (Alberta and Saskatchewan) (Fig. 14–16).

Large deposits of bituminous coal exist in the U.S.S.R., most of them in Siberia (Fig. 14–17). The deposits of the greatest current value are those in the Donets Basin of European Russia, because of its proximity to a long-established concentration of heavy industry and because of the

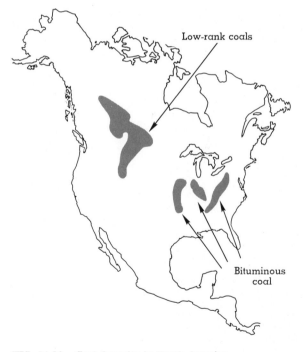

FIG. 14-16. Coal deposits in North America.

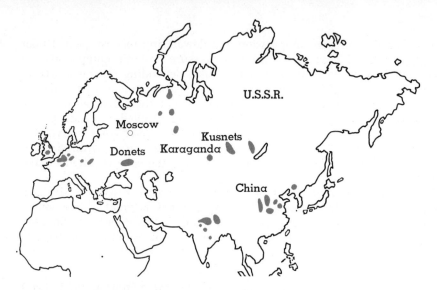

FIG. 14-17. Deposits of bituminous coal in Eurasia.

comparative abundance of coking coal. The Kusnets Basin of south-central Siberia is becoming more significant as time goes on and coal-using areas expand near it. Its reserves are believed to exceed those of the Donets Basin. Smaller coal-mining areas occur elsewhere in the U.S.S.R.

Third in coal reserves in the world, Communist China now ranks second in coal production, which is likely to increase as the country's industrial output rises. Europe ranks next, most production coming from Germany, Great Britain, and Poland (Fig. 14–17). Most of the remaining coal fields in Great Britain are high-cost mining properties, whereas Germany and Poland have great reserves of more accessible deposits.

Africa, Australia, and South America have comparatively insignificant deposits, together totaling less than 3 per cent of the world's reserves.

Petroleum and natural gas are believed to have formed from organic matter of both plant and animal origin. Alteration has proceeded to the point where no recognizable organic structures are now present. These are mobile fluids and are known to have migrated to traps where they have accumulated in large volumes. They occupy space between sediment grains, in openings dissolved in soluble rocks, in openings in biologic sediment such as reefs, and in fractured rocks. The natural site is in sedimentary rocks, and occurrences in igneous or metamorphic rocks are the result of migration from elsewhere. An impervious overlying cap rock is necessary to hold oil and gas in place, and the anticline is the simplest type of trap. Production is achieved by drilling into the porous and permeable reservoir rock, whereupon the fluid flows upward if under pressure or is pumped to the surface.

Production to date and reserves are indicated in Table 14–2, and distribution of economical deposits is shown in Fig. 14–18. Not shown graphically is the much greater number of wells and much longer produc-

TABLE 14–2 PETROLEUM PRODUCTION, 1963, AND ESTIMATED RESERVES, 1964

COUNTRY	PER CENT OF WORLD PRODUCTION	PER CENT OF WORLD RESERVES
U.S.A.	28	10
U.S.S.R.	16	8
Venezuela	13	5
Kuwait	7	Middle East
Saudi Arabia	6	combined, 63
Iran	6	
Iraq	4	

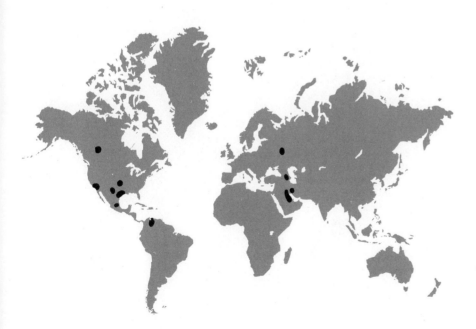

FIG. 14-18. Distribution of major oil-producing areas of the world.

tion time for the United States than for most of the rest of the world. In the Middle East, half a dozen oil wells may produce the equivalent of many hundred wells in Oklahoma and may represent reserves several times larger.

MINERALS

The two groups of mineral products other than fuels are the metallic minerals and the nonmetallic, or so-called industrial, minerals. The requirements for both of these, as for fuels, have increased in variety and quantity throughout man's recent history and accelerating technological development, and will undoubtedly continue to proliferate in predictable and unpredictable ways.

Metallic minerals include both native elements and compounds. Both occur as *ore*, or material which can be economically mined, concentrated, and refined, mixed with a certain amount of *gangue*, or valueless matter. There are many methods of formation of ore deposits, and we shall consider only a few examples as they apply to the metals that are of greatest dollar value in the world today. They apply as well to ores of other metals.

FIG. 14-19. Distribution of major iron-ore deposits of the world.

By the *igneous process*, copper may be concentrated as grains in granitic rocks or in veins in jointed preexisting rocks. Such deposits account for the bulk of the reserves of North America, South America, and Africa (which has the greatest quantity).

By the *weathering process*, metallic ores are concentrated in several ways: (1) The soil-forming process in humid tropical climates results in laterite rich in aluminum compounds; if preserved, it can become bauxite, an ore of aluminum. In this or similar manner were formed the bauxite, reserves in Jamaica, Guyana, and other places in comparatively low latitudes. (2) Iron-rich sedimentary rocks have been residually concentrated in iron through removal of silica by weathering; such deposits are found in many areas. Other iron deposits are of lower grade, as the concentration has not occurred naturally; but they may be economical nonetheless because transportation costs are less and the greater processing expenses can therefore be borne.

Table 14–3 shows statistical information on deposits of iron ore, and the map of Fig. 14–19 indicates districts containing significant deposits. Iron ore, limestone, and coke are necessary in the production of pig iron by the blast furnace. Each is a large-volume, high-tonnage component, so that in the steel industry transportation costs as well as considerations such as purity of deposits and ore and gangue minerals are vital. The availability of water transport provides a distinct economic advantage.

Nonmetallic minerals include a wide range of materials, from such utilitarian products as clay for brick to diamonds for industrial use and feminine adornment, and exotic minerals used in rocket fuels for the space program. In the industrial nations, an accurate gauge of activity is given by the volume of use of halite, sulphur, and calcite, three mineral materials basic to the modern chemical industry.

TABLE 14–3   IRON ORE PRODUCTION, 1963

COUNTRY	PER CENT OF WORLD PRODUCTION	PER CENT OF WORLD RESERVES
U.S.S.R. (1, 2, 3, 4)[a]	32	Extensive reserves are
U.S.A. (5, 6)	16	widely distributed
China (7)	16	throughout the world.
France (8)	8	Brazil (13) has
Canada (9)	6	reserves rivaling other
Sweden (10)	6	resources.
India (11)	5	
Venezuela (12)	3	

[a] Numbers refer to locations on the map of Fig. 14–19.

Tremendous beds of rock salt formed in the past by evaporation of sea water in restricted arms of the oceans are now mined by excavation or by solution in southeastern Michigan, New York State, Kansas, New Mexico, and adjacent areas. In Germany, and in Texas and Louisiana on the Gulf Coast, the nearly flat-lying sedimentary rocks have been intruded by halite from great depths, forming what are known as *salt domes*. On the Gulf Coast the salt is associated with great volumes of sulphur, which are found in the cap rock covering the salt domes. The sulphur is produced by pumping steam and hot water into the salt domes thus liquefying the sulphur and then pumping it to the surface. Elsewhere sulphur is produced from volcanic deposits or is extracted from pyrite, some metal ores, or as a waste product of petroleum refining. Limestone is so widely distributed as a sedimentary rock that few areas can be said to be without it.

WORLD DISTRIBUTION OF POPULATION

Over 80 per cent of the world's population occupies no more than 6 per cent of the land area of the earth. The basic factors accounting for this concentration of population in such a relatively small area are (1) climate, (2) topography, or landforms, (3) mineral resources, and (4) culture and level of technology. The climate and the slope of an area in combination determine the type and intensity of agriculture that can be developed economically. About one-fifth of the earth's land is too cold or has too short a frost-free season for agricultural purposes; over one-quarter is too dry; and almost two-fifths are too rugged. The presence or absence of mineral resources, particularly mineral fuels, iron ore, and other key metallic minerals, greatly influences the possibilities for development of heavy industry and an industrial economy in general. Perhaps the most important factor in explaining the distribution of world population is the uneven level of technology. The way in which the physical environment may be utilized and modified is determined largely by the inventiveness, resourcefulness, and technical knowledge and skills of the people. Societies well advanced in technology may be able to utilize the potential resources of an area and modify or change unfavorable aspects of the physical setting in a fashion that would not be possible for a technologically less-developed people.

The map illustrating the present pattern of distribution of the world's 3.3 billion people shows several areas of sparse population (Fig. 14–20). The northern part of the North American continent, Greenland, and northern Eurasia are very sparsely populated owing to the cold climate (subarctic, tundra, and ice cap), with frost-free seasons too short to permit widespread agricultural development. The excessively dry climatic

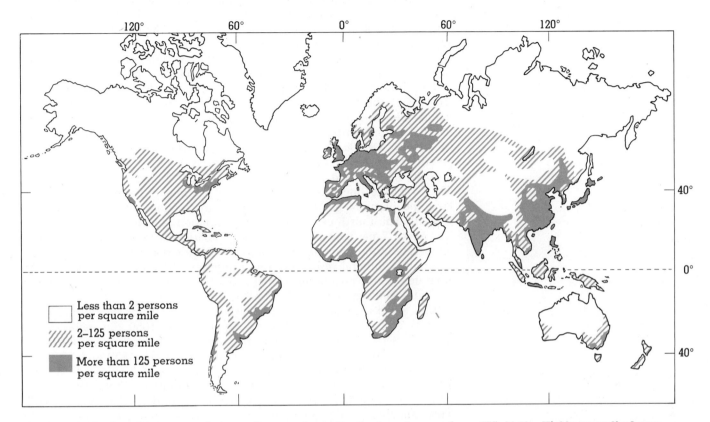

FIG. 14-20. Highly generalized map showing distribution of the world's population.

regions of North Africa, central Australia, southern South America, and southwestern United States and northern Mexico are also sparsely populated, because here precipitation is insufficient for intensive agriculture. Central Eurasia and portions of western North America have rugged highlands of high-slope land and cold climates that are not conducive to the support of dense populations. The sparsity of population in central South America reflects hot and humid tropical rainforest climate, poor drainage, and infertile lateritic soils.

Four major nodes, or centers, of population contain over 85 per cent of the world's peoples. The two largest, both in terms of areal extent and of population, are southern Asia and eastern Asia. Together they account for 57 per cent of the world's population. Within these centers the highest population densities are found in the floodplains and deltas of the larger rivers, such as the Ganges, Indus, Yangtze, Hwang Ho, and Irrawaddy, where fertile alluvial soils, large expanses of low-slope land, and ease of irrigation have been conducive to the development of intensive agriculture. The fertile volcanic soils of Java and the development of industry in Japan are factors that partially account for the very dense populations in those areas. The regions of dense population in southern and eastern Asia lie largely within the areal extent of the A and Ca climates.

The third center of high population comprises Europe and the fertile triangle of the U.S.S.R. About 22 per cent of the world's population inhabits this area of widespread agricultural development and intense industrialization. Europe and the agricultural triangle of the U.S.S.R. are in the Ca, Cb, Cs, Da, and Db climates, all of which have high agricultural potentialities. Much of the land surface is level to rolling, and major deposits of coal and metallic and nonmetallic minerals are widespread. The level of technology is very high, and an industrial economy supporting large populations has developed in many sections of Europe and the U.S.S.R. The fourth center of dense population is in eastern North America. About 7 per cent of the world's population is located here, reflecting temperate humid climate (Ca, Da, Db), large expanses of low-slope land, substantial mineral resources, and a high level of industrialization. Somewhat less-dense population centers are found in western Africa and in coastal Argentina and Brazil (Fig. 14–20).

The world's population is increasing now at a far more rapid rate than at any other time in history. Up to about 1650 the increase was gradual. Between 1650 and 1850 the increase accelerated, averaging 0.4 per cent annually. World population doubled between 1650 and 1850 and doubled again between 1850 and 1950. The rate of increase during the 1850 to 1950 period averaged 0.8 per cent annually. From 1950 to the present time, the increase has averaged slightly under 2 per cent annually. This amounts to a population gain of over 67 million a year. At the present rate of increase, the present population of 3.3 billion will be doubled by 1990.

The rate of population growth differs from one area to another, depending on the pattern of birth and death rates. Table 14–4 shows the three possible combinations—high fertility and low mortality, high fertility and high mortality, and low fertility and low mortality—and the rates of birth, death, and population growth that are characteristic for each. As one would expect, the pattern of high fertility and low mortality is producing the highest rate of population increase—2 to 5 per cent annually. This pattern of growth is now typical of most of Latin America and certain areas of Africa and Asia. A pattern of high fertility and high mortality, with a 1 to 2 per cent annual increase, is characteristic of some of the underdeveloped countries of Asia and Africa. The third pattern of growth is one of low fertility and low mortality, typical of most industrialized and

TABLE 14-4 PATTERNS OF BIRTH AND DEATH RATES

	BIRTH RATE	DEATH RATE	RATE OF POPULATION GROWTH
High fertility, low mortality	40–50 per 1000	10–20 per 1000	2–5%
High fertility, high mortality	40–50 per 1000	25–30 per 1000	1–2%
Low fertility, low mortality	14–25 per 1000	6–12 per 1000	0.5–1.5%

urbanized regions, such as Europe, Anglo-America, the U.S.S.R., Australia and New Zealand, and selected countries in Latin America, Africa, and Asia, such as Japan, Argentina, Uruguay, and the Union of South Africa. Note that although the death rates are the lowest—6 to 12 per 1000 per year—the low birth rates more than balance this, and the population increase is consequently the lowest—½ of 1 per cent to 1½ per cent each year. The present patterns of population growth indicate that by the year 2000 about 62 per cent of the world's people will be in Asia, mostly the southern and eastern portions; 10 per cent in Africa; 14 per cent in Europe and the U.S.S.R.; 9 per cent in Latin America; 4 per cent in North America; and less than 1 per cent in Oceania.

An understanding of the physical environment is pertinent today to the problem of rapidly exploding population and the attempts to raise the standard of living in underdeveloped countries. Only about 8 per cent of the land area of the earth is suitable for cultivation from the standpoint of climate and slope. Presently, two-thirds of the world's people do not have an adequate or well-balanced diet, and the relatively small amount of arable land must produce the foods and fibers for a much larger population in the future. Although the most accessible and most economical deposits of many key mineral resources have been depleted, the demand for these resources will become even greater than at present. It is essential that our physical environment be used wisely and that knowledge of modern farming and mining techniques be widely disseminated if the problem of adequate resources for the world's peoples is to have any chance of being solved. It is also vital that investigations already undertaken be continued and expanded to determine how the oceans can be better utilized, the possible future role of synthetic foods and vitamins, the feasibility of producing foods and fibers by hydroponics, possible means of exploiting mineral deposits not now economical, and possibilities for substituting other materials for minerals and fuels nearing depletion.

SUGGESTED REFERENCES

Hoy, J. B.: *Man and the Earth*, Prentice-Hall, Inc., Englewood Cliffs, N.J., 1967.

James, P. E.: *A Geography of Man*, Blaisdell Publishing Company, Waltham, Mass., 1966.

Kendall, H. M., R. M. Glendinning, and C. H. MacFadden: *Introduction to Geography*, Harcourt, Brace and World, Inc., New York, 1967.

McIntyre, M. P.: *Physical Geography*, Ronald Press Company, New York, 1966.

Mudd, S. (ed.): *The Population Crisis and the Use of World Resources*, Indiana University Press, Bloomington, Ind., 1964.

Murphey, R.: *An Introduction to Geography*, Rand McNally and Company, Chicago, 1966.

Strahler, A. N.: *Introduction to Physical Geography*, John Wiley & Sons, Inc., New York, 1965.

Trewartha, G. T., A. H. Robinson, and E. H. Hammond: *Elements of Geography*, McGraw-Hill Book Company, Inc., New York, 1967.

Van Riper, J. E.: *Man's Physical World*, McGraw-Hill Book Company, Inc., New York, 1962.

TOPOGRAPHIC MAPS
AND AERIAL PHOTOGRAPHS

Maps are scientific documents that organize and present graphically information illustrating the spatial distribution of various types of phenomena existing on the earth's surface. Inasmuch as maps are small-scale substitutes for actual segments of the earth, they cannot illustrate each and every aspect of the surface. Certain features must be selected to be represented, and these features must then be generalized and categorized. The scale of the map, selection of phenomena to be shown, and the manner of classifying these phenomena determine how closely the map approximates reality. An understanding of maps and the ability to use them skillfully are attributes of every educated person and are essential for the student in geography, geology, and earth science in general.

There are many kinds of maps: some cover the entire world, some a hemisphere, and some a continent. Maps of large earth areas are called small-scale maps; maps of small areas such as townships, cities, or counties, are large-scale maps. The scale of a map may be expressed as a fraction, called the representative fraction, or RF for short. For example, an RF of 1:10,000 means that one unit on the map represents 10,000 of the same units on the earth.

TOPOGRAPHIC MAPS

In the study of landforms or cultural features, a specific type of large-scale map called a *topographic quadrangle,* published by the U.S. Geological Survey, is commonly used. A portion of a topographic quadrangle is shown on the inside back cover. These maps are usually at the scale of 1:62,500 (approximately 1 inch = 1 mile) or 1:24,000 (approximately 2½ inches = 1 mile). However, other larger or smaller scales may be used. They are printed in three to five colors. Cultural or man-made features such as roads, houses, urban areas, boundaries, etc., are in black and, to a lesser degree, in red; all water features, both man-made and natural, are in blue; symbols indicating the elevation of land surfaces are in brown; and, on some maps, timber or woodland areas are shown in green.

Topographic maps use contour lines to illustrate relief. A contour line runs through points of equal elevation above sea level and has a value in feet or meters. The vertical distance between consecutive contour lines is called the contour interval and varies from one map to another. It may be hundreds of feet in mountainous areas of high relief or as little as 5 feet or even less in areas of relatively level topography. Also, the contour interval usually becomes larger as the scale of the map becomes smaller. Contour lines provide reasonably accurate information as to the elevation of any point within the boundary of the map and the outline of specific landforms, as well as showing the pattern or arrangement of groups of landforms. Topographic maps are essential tools for accurately studying areas of the earth's surface, for it is obviously not possible to bring large segments of the world into the laboratory for analysis.

The location of a place or an area on a topographic map may be described in terms of longitude and latitude. Lines of latitude (parallels), and lines of longitude (meridians), comprise a coordinate system superimposed on the earth's surface. Lines of latitude run east-west and are used to measure distances from the equator (latitude 0°) to the South and North Poles (90° south and 90° north, respectively). Lines of longitude run north-south through the poles and are used in measuring distances east and west from the prime meridian, which passes through the Greenwich Observatory in London, England. The degrees of longitude are numbered from the prime meridian 180° both east and west. Each degree of longitude and latitude is divided into 60 minutes (60′), and each minute into 60 seconds (60″). The longitude and latitude covered by a topographic map are always shown in the margin, and the most commonly used topographic maps cover an area of either 15′ of latitude and longitude or 7½′ of latitude and longitude.

Another way in which areas or specific points may be located is by using the range and township system, common to most of the United States west of the eastern-seaboard states (Fig. I–1). This system employs selected lines of latitude called base lines and selected lines of longitude called principal meridians. The land is divided into essentially square units by means of ranges and townships. Townships are 6-mile-wide strips of land trending east and west, and each is numbered north or south from a specific base line. Each township is divided into a series of ranges by north-south lines, spaced 6 miles apart, and numbered from a specific principal meridian. Because meridians converge on the poles, not all adjacent meridians can be 6 miles apart. Corrections for the convergence are made at intervals, and not all townships are perfectly square. Normally, each township is further divided into 36 numbered sections, each 1 square mile in size (Fig. I–2). Each section may be divided into quarters, and so on, if more precise location is necessary (Fig. I–3). Point A in Figs. I–1 to I–3 may then be located in the SE¼ of the SW¼ of Section 8 T2S, R1E.

Direction is shown on a topographic map by arrows indicating true and magnetic north. The magnetic North or South Poles and the geographic poles do not coincide, so that in only a few areas is the magnetic needle parallel to a line of longitude. The angle between magnetic and true north varies considerably from place to place, and this variation is referred to as *declination*. The Agonic Line running through the central United States

T2N
T1N
T1S
T2S
Base
line
Principal
meridian
•A
R2W
R1W
R1E
R2E
0
6
12
Miles

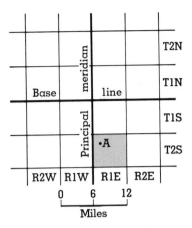

FIG. I-1. Township and range system.

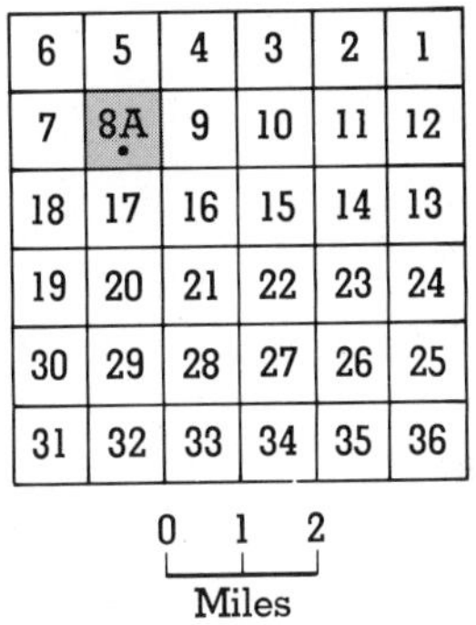

6 5 4 3 2 1
7 8A 9 10 11 12
18 17 16 15 14 13
19 20 21 22 23 24
30 29 28 27 26 25
31 32 33 34 35 36
0 1 2
Miles

FIG. I-2. The usual system of numbering sections within a township.

5
7
8
9
A•
17
0 1320 2640 5280
Feet

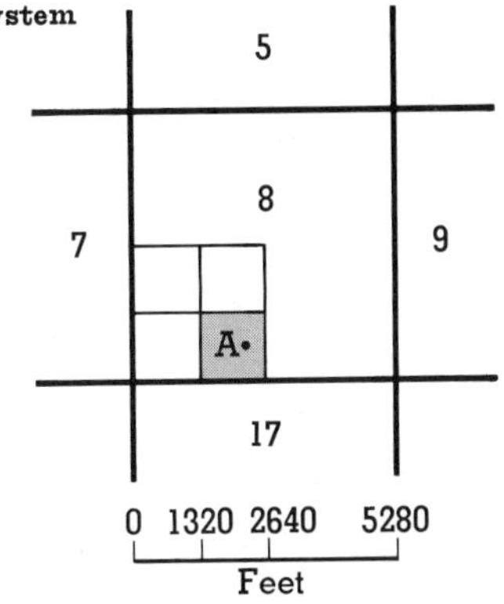

FIG. I-3. A location described by quarter sections.

from about the eastern seaboard of Florida to Lake Superior has 0° declination (Fig. I–4). The declination of the area covered is shown in the margin of each topographic map.

AERIAL PHOTOGRAPHS

The aerial photograph is another valuable tool for accurate study of landscape features. Inasmuch as it is a photograph of an actual part of the earth, it shows many features such as fields, vegetation, and soils that do not normally appear on a topographic map. The most commonly used aerial photograph is the vertical photograph, which portrays accurately the land in the central portion, directly beneath the camera. Near the edges of the photograph, distortion occurs because of the inclined view of the camera. Vertical photographs may be taken in a series in such a way that the central portions of each may be joined together to form a large photo-map. The photographs are taken in a series providing considerable overlap, so that any given feature will appear two or more times. By means of these overlapping photographs and a stereoscope, the dimensions of depth may be visualized.

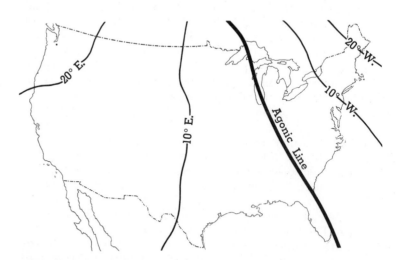

FIG. I-4. Lines of equal magnetic declination in the United States.

INDEX

INDEX

Cover photo by Fotis Studio

Set in Linotype Primer
Format by Frances Torbert Tilley
Composition by The Haddon Craftsmen
Printed by The Murray Printing Co.
Manufactured by The Haddon Craftsmen

70 71 7 6 5 4 3